THE CALCULATION OF MOLECULAR ORBITALS

The Calculation of Molecular Orbitals

JOHN C. SLATER
Institute Professor Emeritus
Massachusetts Institute of Technology

Graduate Research Professor of Physics and Chemistry
University of Florida

A WILEY-INTERSCIENCE PUBLICATION

JOHN WILEY & SONS, New York · Chichester · Brisbane · Toronto

CHEMISTRY

Copyright © 1979 by John Wiley & Sons, Inc.

All rights reserved. Published simultaneously in Canada.

Reproduction or translation of any part of this work beyond that permitted by Sections 107 or 108 of the 1976 United States Copyright Act without the permission of the copyright owner is unlawful. Requests for permission or further information should be addressed to the Permissions Department, John Wiley & Sons, Inc.

Library of Congress Cataloging in Publication Data

Slater, John Clarke, 1900–1976.
   The calculation of molecular orbitals.

   "A Wiley-Interscience publication."
   Bibliography: p.
   Includes index.
   1. Molecular orbitals. I. Title.

QD461.S49   1978   541'.22   78-323
ISBN 0-471-03181-X

Printed in the United States of America

10 9 8 7 6 5 4 3 2 1

Foreword

At the time of his sudden death in 1976 the manuscript of this book was found among Professor Slater's effects. It is essentially complete, having been finished just a few days before. The only addition that has been made is a list of references relevant to the discussion in the text.

This work represents the last in a series of books by Professor Slater, on the quantum theory of matter, consisting of seven volumes, published from 1960 to 1975. This monumental work covers all aspects of the modern theory of atoms, molecules, and solids. The present volume could be considered an addendum to this series, in that it is directed toward an unsolved problem of the quantum theory of matter, that is, how to solve exactly the self-consistent field problem.

In the last ten years of his life Professor Slater had been concerned with the local density form of the self-consistent field. His work, together with that of many collaborators, had resulted in the $X\alpha$ method. Most of his publications during this time were concerned with the development of this method and its application to a wide range of problems. The usefulness of this model has been remarkable, and it has greatly helped in the understanding of complicated systems, for which more sophisticated methods are not applicable because of the limitations of present-day computers. However, the equations resulting from the $X\alpha$ model have not yet been solved exactly. The applications so far have all involved approximations such as that of the "muffin-tin" potential, perturbation theory, the use of a limited basis set, and many other numerical methods. These approximations are not so severe as to invalidate the calculated results for many of the properties of molecules and solids. However, in certain cases, they have proved to be inadequate.

It was Professor Slater's purpose, in this book, to suggest a different scheme of solution of the equations. This scheme is essentially a revival of the cellular method, invented forty years ago but presented in the light of recent developments. The method as presented here is not complete in all details but is simply an outline of the directions in which Professor Slater thought the research should proceed. It is indeed a great loss that he will not be around to continue the work. However, it is hoped that the publication of this book will encourage others to follow along the same lines and perhaps to complete the research to the end which Professor Slater intended.

JOHN W. D. CONNOLLY

Washington, D. C.
January 1979

Contents

1. Molecular Orbitals and the Self-Consistent Field, **1**
2. The $X\alpha$ Method, **8**
3. Schrödinger's Equation for the Spherical Problem, **13**
4. Power Series Expansions in the Spherical Problem, **20**
5. Determination of the Self-Consistent Field for Atoms, **35**
6. The Total Energy of An Atom, **43**
7. Exitations and the Transition State, **49**
8. Spin Polarization and the Unrestricted Self-Consistent Field, **53**
9. The Cellular and Muffin-Tin Methods, **55**
10. The Multiple-Scattering Method, Weak Interaction, **63**
11. Qualitative Nature of Molecular Orbitals, **69**
12. The Cellular Method-Solution of Schrödinger's Equation, **81**
13. Self-Consistent Field for Molecules, **88**
14. The Nonspherical Potential, **92**

Notes, **99**

Bibliography, **101**

Index, **103**

1 Molecular orbitals and the self-consistent field

The concept of molecular orbitals is inextricably tied up with the idea of the self-consistent field. In a molecular system consisting of N electrons and many nuclei, the latter being assumed to be at rest, each electron will really move in the electric field produced by all $N - 1$ other electrons and by all the nuclei. If N is greater than a very small number, this forms an impossibly difficult problem for rigorous solution in quantum mechanics, as it would be impossibly difficult in classical Newtonian mechanics. One recalls the very great difficulty of even the celebrated classical three-body problem, $N = 3$, in celestial mechanics.

Workers in the electronic problem of space charge in a vacuum tube encountered this difficulty many years ago, and overcame it by a very obvious device: they assumed that the effect of all $N - 1$ other electrons could be approximated by averaging the density of these other electrons over their complex motions and finding the electric field arising from this continuous averaged density. They studied the motion of the one remaining electron in the field of these $N - 1$ averaged charges. Then they demanded that the paths of the electrons, so computed, should lead to the same charge density that was assumed in the first place as that of the space charge.

The electric field so set up, produced by the electrons in their averaged motions and by any electrodes present (in the electronics problem) or by any nuclei present (in the molecule), is what we now call a self-consistent field. The name was introduced by D. R. Hartree in 1928. He was studying the electrons in an atom, moving around its nucleus. The same concept was applied at the same time by F. Hund, R. S. Mulliken, and J. E. Lennard-Jones to the molecular problem, and it formed the basis of theories of electrons in crystals set up around the same time by W. Heisenberg, F. Bloch, L. Brillouin, A. H. Wilson, and many others. We now apply the terminology of the self-consistent field to all these problems. The individual electrons then move in a fixed external field, and their wave functions, as found from the solution of Schrödinger's equation, are called orbitals—atomic orbitals for the atomic case, molecular orbitals for molecules or crystals.

The effect of this approximation is to reduce the many-body problem from a $3N$-dimensional one, in the three coordinates of all N electrons, to N separate three-dimensional problems, an enormous simplification. The Schrödinger equation for one of the molecular orbitals is still a difficult one. For the atom, Hartree's problem, it is quite simple, on account of the spherical symmetry. But in the molecule or crystal we no longer have this symmetry, except in the neighborhood of one of the nuclei. A large part of our effort in this volume must go to examining adequate ways to handle the nonspherical problem of Schrödinger's equation in the molecule, making full use of our experience in solving the spherical atomic problem. But we must also investigate the total energy of the many-electron problem, as we compute it from the atomic orbitals determined self-consistently. We shall first look into this question of the interrelationship of the three-dimensional and $3N$-dimensional problems, and shall find that an approximation, the so-called $X\alpha$ approximation, can be set up to overcome the larger part of the difficulty of the $3N$-dimensional problem. Then we go on to apply these methods to actual molecular calculations.

We start with Hartree's ideas. He was considering an atom or ion of N electrons, surrounding a nucleus at rest, with atomic number Z. Each electron, of course, is acted on electrostatically by the nucleus and all $N - 1$ of the other electrons. He wished to replace the very complicated field exerted by the other electrons, which depends on just where these electrons are, by a single potential arising from the average positions of the other electrons. He then solved the Schrödinger equation for wave functions of a single electron moving subject to this averaged potential. If u_i is such a wave function or atomic orbital, normalized so that the integral over all space of $u_i^* u_i$ is unity, he assumed that N of these wave functions represented occupied states in the atom and that $u_i^* u_i$ represented the magnitude of the charge density (in units of the electronic charge) of the ith electron. The total charge density of all electrons would then be $-\sum(i) u_i^* u_i$, where the summation goes over the occupied orbitals and where we have used the minus sign to indicate that the electrons have a negative charge. We find it convenient to assign occupation numbers n_i to the states, unity for an occupied state, zero for an empty one. Then the total electronic charge density is

$$\rho = -\sum(i) n_i u_i^* u_i \qquad (1\text{-}1)$$

where now the summation can go over all orbitals, both occupied and empty.

Hartree next set up the potential in which the ith electron moves. We use atomic units: the rydberg as a unit of energy, (although Hartree used 2 rydbergs as a unit, ordinarily called a hartree), and a unit of distance, now called the bohr, equal to the radius of the $1s$ orbit of hydrogen in Bohr's atomic theory. In these units the potential arising from the nucleus, at distance r, is $2Z/r$. The potential arising from the charge located in all volume elements dv_2, at a point 1 distant from such a volume element by a distance

1 MOLECULAR ORBITALS AND THE SELF-CONSISTENT FIELD

r_{12}, is $\int \rho(2)(2/r_{12})\,dv_2$, where the integral extends over all space and where $\rho(2)$ is the charge density at point 2. However, Hartree naturally assumed that a given electron could not act on itself. Hence for an electron in the ith orbital, he omitted the quantity $u_i^* u_i$ from the charge density as given in Equation 1-1. We can describe the situation by saying that the potential acting on the ith electron at position 1 is

$$V_i(1) = V_N(1) + V_e(1) + V_{Xi}(1) \tag{1-2}$$

where

$$V_N(1) = \frac{2Z}{r_1}, \quad V_e(1) = \int \rho(2)\left(\frac{2}{r_{12}}\right) dv_2,$$

$$V_{Xi}(1) = \int u_i^*(2) u_i(2) \left(\frac{2}{r_{12}}\right) dv_2$$

Thus V_N is the nuclear potential, V_e the electronic potential arising from all electrons, and V_{Xi} the correction term arising because the electron in one orbital does not act on itself.

Hartree then assumed that the electron moving in the ith orbital had a Schrödinger equation

$$-\nabla^2 u_i(1) - V_i u_i(1) = \varepsilon_i u_i(1) \tag{1-3}$$

where $-\nabla^2$ is the kinetic energy in our atomic units, $-V_i$ is the potential energy of the negative electron in the potential V_i of Equation 1-2, and ε_i is the one-electron energy. For the spherical symmetry found in the atom it is easy to solve Equation 1-3, as we show in detail in Chapters 3–5. The solutions, in spherical polar coordinates, are products of spherical harmonics of the angles and a function of the radius vector r. This radial function has an ordinary differential equation which can be easily solved by numerical methods. For certain discrete (negative) energies—the eigenvalues of the problem—we find functions—the eigenfunctions—that are regular both at the nucleus and at infinity. Hartree demanded for self-consistency that these eigenfunctions, normalized, should be identical with the functions u_i met in Equation 1-1.

This would not automatically happen, and Hartree devised a method of successive approximations, or iteration, to secure functions which had this property. Namely, he took the u_i's resulting from solving Equation 1-3 at one stage of the process and substituted them in Equation 1-1 for setting up the next stage of iteration. After a number of iterations, he found that the initial and final u_i's were identical. He gave the name self-consistent to the resulting field, potential, and wave functions. He found that the self-consistent charge density from Equation 1-1 gave a very good approximation to the experimentally determined charge density in the atom. Furthermore, the difference between the energy values of an occupied state and an empty state gave a good approximation to the experimentally determined excitation energies. Ordinarily in the ground state of an atom it was found that the eigenfunctions of lowest energy should be occupied, the higher ones empty, although occasional cases did not fit in with this general rule.

At the same time Hartree was doing this work on atoms, we have mentioned earlier that Hund, Mulliken, and Lennard-Jones were considering simple molecules from essentially the same point of view, although the two sets of workers were at first independent of each other. The only difference in the formulation is that the nuclear potential $V_N = 2Z/r_1$ of Equation 1-2 had to be replaced by a sum of such terms for the potential arising from all nuclei in the molecule. The equivalent Schrödinger equation, however, was felt almost impossibly difficult to solve directly, because it was not spherically symmetrical. Thus the results of this method of investigation of molecules were used only as a qualitative procedure until after World War II.

In 1930 it occurred to several workers that it ought to be possible to set up a many-electron wave function from the atomic orbitals u_i and to apply the variation method of quantum mechanics to this wave function. It is a fundamental principle of wave mechanics that if we have an approximate wave function and compute from it the average value of the many-electron Hamiltonian H over it, varying the approximate wave function to minimize the average value, the result will represent the closest approximation to the true wave function we can attain using the set of functions considered, and the average Hamiltonian must lie higher than the ground-state energy. If we set up a product function, $u_1(1)u_2(2) \cdots u_N(N)$, where $u_1 \cdots u_N$ represent the N occupied orbitals and $(1), \ldots, (N)$ represent the coordinates of the N electrons, the product of this function and its complex conjugate is a product of quantities $u_i^*(i)u_i(i)$ for the various electrons. This product would indicate that the electrons move independently of each other, which is at the foundation of Hartree's idea of self-consistency. The many-electron Hamiltonian for an atomic or molecular system, in the atomic units we are using, is

$$-\sum(i)\nabla_i^2 - \sum(i,a)\frac{2Z_a}{r_{ia}} + \sum(\text{pairs } ij, i \neq j)\frac{2}{r_{ij}} + \sum(\text{pairs } ab, a \neq b)\frac{2Z_aZ_b}{r_{ab}}$$

(1-4)

Here the indices i refer to the electrons; the indices a refer to the nuclei; r_{ia} is the distance between the ith electron and the nucleus a; r_{ij} is the distance between electrons i and j, and r_{ab} is the distance between nuclei a and b. In this case, as in all others in the text, the nuclei are assumed to be at rest.

When this Hamiltonian was allowed to operate on the product wave function $u_1(1)\cdots u_N(N)$, multiplied by the conjugate of the wave function, and integrated over all values of the $3N$-dimensional space of the $3N$ electronic coordinates, one had an expression for the total energy of the atom or molecule. If one of the orbitals was varied and the change in energy was computed, it was found that the energy minimum came precisely when the u's satisfied Hartree's Equation 1-3. This thus formed a theoretical basis for the procedure Hartree had arrived at by intuition.

In the meantime, it had been found that the antisymmetry of the wave function when the coordinates (and spins) of two electrons were interchanged was an expression of Pauli's exclusion principle. One had to enlarge the meaning of the orbitals u_i to include dependence on spins, so that for each electron one could have both a spin-up and a spin-down orbital, hence called a spin orbital. To secure the required antisymmetry, it was necessary to write the many-electron wave function not in the form of a product $u_1(1)\cdots u_N(N)$ but in the form of a determinant,

$$(N!)^{-1}\begin{vmatrix} u_1(1) & u_1(2) & \cdots & u_1(N) \\ u_2(1) & u_2(2) & \cdots & u_2(N) \\ \multicolumn{4}{c}{\dotfill} \\ u_N(1) & u_N(2) & \cdots & u_N(N) \end{vmatrix} \tag{1-5}$$

where the factor $(N!)^{-1}$ gives a normalized function, if the spin orbitals u_i are orthonormal.

In this case, one can still vary one of the spin orbitals u_i to minimize the total energy. This total energy has the form

$$\langle EHF\rangle = -\sum(i)n_i\int u_i^*\nabla^2 u_i\,dv - \sum(i)n_i\int u_i^* V_N u_i\,dv - \tfrac{1}{2}\sum(i)n_i\int u_i^* V_e u_i\,dv$$

$$-\tfrac{1}{2}\sum(i,j)n_i n_j\delta(m_{si}m_{sj})\int u_i^*(1)u_j^*(2)\left(\frac{2}{r_{12}}\right)u_j(1)u_i(2)\,dv_1\,dv_2$$

$$+\sum(\text{pairs }ab,\,a\ne b)\frac{2Z_a Z_b}{r_{ab}} \tag{1-6}$$

where $V_N = \sum(a)2Z_a/r_{1a}$, V_e is given in Equation 1-2, and $\delta(m_{si}m_{sj})$ indicates that one includes only those pairs of occupied orbitals i, j for which the spin quantum numbers m_{si} or m_{sj} are the same or pairs of spin orbitals with parallel spins. When one varies one of the spin orbitals to minimize the total energy, the resulting one-electron Schrödinger equation is

$$-\nabla^2 u_i - (V_N + V_e)u_i - \sum(j)\,\delta(m_{si}m_{sj})\left[\int u_j^*(2)\left(\frac{2}{r_{12}}\right)u_i(2)\,dv_2\right]u_j = \varepsilon_i u_i \quad (1\text{-}7)$$

The self-consistent-field method using Equations 1-6 and 1-7 is called the Hartree–Fock method. Hartree and Fock both were able to develop the method into a practical form for solving atomic problems, and the results proved to be slightly closer to experiment than those obtained by the original Hartree method. The last term of Equations 1-6 and 1-7 is obviously much more complicated than the corresponding one in Equations 1-2 and 1-3. In connection with the Hartree–Fock equations it is generally referred to as the exchange term, which is why we have used the subscript X to refer to it. Its dependence on spin has led to many applications in magnetic problems. But its existence in the Hartree–Fock equations complicated the molecular problem so much that it seemed quite out of the question to make any straightforward calculation of molecular orbitals using the Hartree–Fock method, except for the very simplest molecules (in practice, for H_2).

However, the interpretation of the Hartree–Fock method as one for minimizing the energy of Equation 1-6, with the Hamiltonian of Equation 1-4, suggested quite a different approach to the solution of the molecular orbital problem, which has been widely developed since World War II. This was based on the so-called LCAO, (linear combination of atomic orbitals) method, which had been used in a qualitative way since the earliest days of molecular orbitals. Approximate ways of computing molecular orbitals had shown that a linear combination of atomic orbitals, located on the various nuclei taking part in a chemical bond, gave quite a good first approximation. Furthermore, it had been shown that quite good approximations to atomic orbitals could be set up by use of functions $\exp(-ar)r^n$ times a spherical harmonic of the angles. Linear combinations of such functions on each of the atoms concerned were then substituted for the u_i's of Equation 1-6, with coefficients to be determined so as to minimize the energy. C. C. J. Roothaan, in 1951, formulated the required equations for the coefficients, which have been widely used. A great deal of effort was put into the determination of the

1 MOLECULAR ORBITALS AND THE SELF-CONSISTENT FIELD

best constants a to use in the various basis functions $\exp(-ar)r^n$. When sufficiently large basis sets are used, very close approaches to Hartree–Fock molecular orbitals can be set up in this way.

It is, however, extremely demanding in the amount of computer time and capacity required. For this reason it was only in the 1960s that really good results began to be obtained by use of this method. An important improvement in technique arose from the discovery of S. F. Boys, during the decade of the 1950s, that if the orbitals were made up out of Gaussian functions of the form $\exp(-ar^2)r^n$ instead of the functions $\exp(-ar)r^n$, the necessary integrals were much simpler to compute. Most of the present work is being done with the use of the Gaussian functions. But when one multiplies the exchange term of Equation 1-6 by the complex conjugate of one of the orbitals and integrates over dv_1, as is required to get the exchange terms in the total energy, one is dealing with a product of four orbitals. Each term leads to what is called a four-center integral, since the atomic orbitals of each of the molecular orbitals can be located on four separate nuclei. The experts in the use of this method, for instance E. Clementi, were speaking of literally billions of integrals which had to be computed and combined to get the complete result. For a large molecule, only the very largest computers sufficed, and the computing time required was quite colossal. As the 1960s led into the 1970s, it became quite obvious that this method was bound to lead to eventual trouble, since one was reaching the limit of what any computer in sight could do.

In the meantime, solid-state theory had been proceeding along quite a different direction. E. Wigner and F. Seitz, in 1933, had suggested what is now called a cellular method of handling the problem of computing the molecular orbitals of a simple crystal. These orbitals have eigenvalues ε which form continuous bands, for which reason this type of theory is known as energy band theory. Furthermore, Wigner and Seitz had made very valuable observations about the simplification of the exchange terms. They had used corresponding terms developed earlier by Dirac and Bloch for studying the magnetic properties of an electron gas.

Unfortunately, the quantum chemical theorists were almost completely unaware of these developments. However, the author, interested in both the molecular and the solid-state theory, realized from 1933 that the general methods introduced by Wigner and Seitz and developed during the 1930s into practical methods for calculating energy bands should be equally adaptable to the molecular problem. For a variety of reasons, it has taken up to the present decade to get these methods into a form adaptable for molecular calculations.

In the MS-$X\alpha$ method of K. H. Johnson and F. C. Smith, Jr., (MS refers to multiple scattering; $X\alpha$ to an exchange term with a parameter alpha), we have a method of calculation going back in spirit to the original calculations of Hartree, using a straightforward solution of the one-electron Schrödinger equation, rather than the LCAO schemes which lead to the great computational difficulties. Results of these newer methods, on quite complicated molecules, show accuracy which is in most cases better than the best Hartree–Fock calculations done by the LCAO method, with computer times in many cases a thousand times less than for the conventional methods. There are some features in which this MS-$X\alpha$ method is not completely rigorous and satisfactory. But outgrowths of this method, which are described in the following chapters, give promise of overcoming the unsatisfactory features, while retaining the computational advantages of the method.

2 The X alpha method

At the end of the preceding chapter it was mentioned that Wigner and Seitz had made use of earlier work by Dirac and Bloch in setting up a simplified form for the exchange term. Dirac and Bloch had been considering a perfect gas composed of N electrons. The one-electron wave function u_i of an electron in a perfect gas is of the nature of a plane wave, expressed in exponential form as $\exp(ik \cdot r)$, where k is a so-called wave vector, r the radius vector. One can set up the determinantal many-electron function for N electrons in the form of Equation 1-5. If $N\uparrow$ of the electrons have spins pointing up and $N\downarrow$ have spins pointing down, the determinant factors into two terms, one a determinant formed from the $N\uparrow$ electrons with spin pointing up, the other formed from the electrons with spin pointing down.

If we now fix our attention on an electron with spin up, at a given position in space, we can answer from the wave function the question, what is the

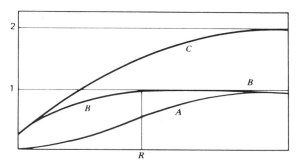

Figure 2.1 Density of charge near an electron, plotted against distance from a given electron. Curve A for another electron of same spin, B of opposite spin, C for both spins combined. One unit of density represents maximum allowable value for electron of one spin. Integrated deficiency of charge, for curves A and C, 1 electron unit; for B, zero. From Slater, *Rev. Mod. Phys.*, 6, 209 (1934).

density of other electrons of spin up in the neighborhood of the one electron at the given position? We find that at a large distance from the fixed electron, the density is just the average value we should expect to find in the perfect gas. Near the given electron, however, there is a deficiency of charge, as indicated in Figure 2.1. The density of other electrons of the same spin falls to zero at the position of the fixed electron, and the integrated deficiency of electronic charge amounts to exactly one electron. This electron deficiency is called the Fermi hole.

The meaning of the Fermi hole is obvious. There are only $N\uparrow - 1$ other electrons of spin up, aside from the electron at the fixed position. Thus the Fermi hole represents the sphere of influence, so to speak, from which one electron is removed, so that the remaining charge density will integrate to $N\uparrow - 1$ electrons. There is no corresponding Fermi hole for electrons of spin down, near an electron of spin up. There are $N\downarrow$ electrons of spin down, and to the approximation in which the electrons are represented by a perfect gas, they have a uniform unperturbed density in the neighborhood of an electron of spin up. Wigner and Seitz, however, realized that an electron of spin up would repel an electron of spin down electrostatically, and that this would tend to keep them apart. They concluded that the probability of finding an electron of spin down near an electron of spin up would be given by a curve similar to that of curve B in Figure 2.1, whereas the probability of finding the electron of spin up is given by curve A. This additional effect, a result of electrostatics rather than of the antisymmetry of the wave function, is generally called the correlation effect. Its importance is rather small compared to the effect of the Fermi hole.

Dirac and Bloch studied the effect of the Fermi hole on the total energy of the electron gas. They pictured a free electron gas as consisting of N electrons, which, of course, would carry a very large negative charge, plus a uniformly distributed positive charge, just enough to cancel the negative charge. This positive charge was supposed to take the place of the nuclei. Then an electron would feel no electrostatic effect whatever, since all electrostatic charge was neutralized, except for its interaction with the positive charge left unneutralized because an electron was removed from the Fermi hole. In other words, the electron had a potential energy as if it were at the center of a positive charge distribution, arising from a single positive charge distributed as in the Fermi hole.

We can use a simple argument to get the dimensional form of the resulting energy term. Let us suppose that the density of electrons of spin up is ρ. If we replaced the Fermi hole by a sphere, such that within this sphere of radius r there was no electronic charge of spin up, whereas outside the sphere the density was $\rho\uparrow$, we can calculate the radius of the sphere. The volume of the sphere is $4/3\,\pi r^3$, so that we have the condition that this volume times the density of charge must equal one electron. That is,

$$\frac{4}{3}\pi r^3 \rho\uparrow = -1 \tag{2-1}$$

where the minus sign comes in because we are treating the electronic density as negative. Hence we have $r = (-3/4\pi\rho\uparrow)^{1/3}$. But the potential energy of an electron at the center of a uniformly charged distribution of this sort can be shown by electrostatics to be $-3/r$ in our atomic units. Thus we should find that the energy of the electron was

$$\text{energy} = -3\left(\frac{-4\pi\rho\uparrow}{3}\right)^{1/3} \tag{2-2}$$

This argument is oversimplified, because, actually, the Fermi hole is like that shown in Figure 2.1, rather than being a spherical hole with a sharp edge. But the simple argument has shown correctly that the energy arising from the Fermi hole should be proportional to the $\frac{1}{3}$ power of the electron density of spin up. The only correction that is needed is in the value of the constant coefficient. It is this term that takes the place of the V_{Xi} of Equation 1-2 or

2 THE X ALPHA METHOD

the more complicated form found for the Hartree–Fock method in Equation 1-6. Shortly we shall introduce a parameter α into this term to serve as a correction to a simple derivation such as we have just given. Then we arrive at the quantity $V_{X\alpha}$, which is to be used in place of V_{Xi} in Equation 1-2. But the situation is, in fact, more complicated than one might think, and we shall now explain this complication.

The complication arises because the exchange term, like Equation 2-2 but with a revised coefficient, appears in two different places in the theory. First, it appears in the one-electron Schrödinger equation, like Equation 1-3. But second, it appears in the expression for the total energy of the molecular system. In the Hartree–Fock method, the corresponding term, given in Equations 1-6 and 1-7, also appears in both places, with closely related coefficients in each. In the $X\alpha$ method, the coefficient multiplying the expression in the one-electron equation is differently related to that for total energy. The author overlooked this fact in the first work with the $X\alpha$ method, in 1951. It was pointed out in 1954 by Gaspar, and later in 1965 by Kohn and Sham. When the fact was realized, it seemed most convenient to introduce a factor α into the energy expressions in a way we shall now make precise.

Let us first write down the expression for the total energy of the molecule in the $X\alpha$ method. We call this $\langle EX\alpha \rangle$. It is

$$\langle EX\alpha \rangle = -\sum(i)n_i \int u_i^* \nabla^2 u_i \, dv - \sum(i)n_i \int u_i^* \sum(a) \left(\frac{2Z_a}{r_{1a}}\right) u_i \, dv$$

$$+ \frac{1}{2} \int \rho(1)\rho(2) \left(\frac{2}{r_{12}}\right) dv_1 \, dv_2$$

$$- \frac{9}{2}\alpha \left(\frac{3}{4\pi}\right)^{1/3} \int \{[-\rho\uparrow(1)]^{4/3} + [-\rho\downarrow(1)]^{4/3}\} \, dv_1$$

$$+ \sum(\text{pairs } ab, a \neq b) \frac{2Z_a Z_b}{r_{ab}} \tag{2-3}$$

Here n_i, u_i have the same significance as in Equation 1-1, and ρ, Z_a, r_{1a}, r_{12}, r_{ab} are as in Equations 1-2 and 1-4. The quantities $[-\rho\uparrow(1)]^{4/3}$, $[-\rho\downarrow(1)]^{4/3}$ arise from the product of the quantity $[-\rho\uparrow(1)]^{1/3}$ or $[-\rho\downarrow(1)]^{1/3}$ of Equation 2-2 and the corresponding charge density. The value of the exchange term, the term in α in Equation 2-3 that corresponds to the free-electron gas, is given by $\alpha = 2/3$. This is often referred to as the Kohn–Sham value.

We can now vary one of the spin orbitals, u_i^* to minimize the energy, treating α as a constant, and obtain a one-electron equation. This is

$$-\nabla_1^2 u_i(1) - \sum(a)\left(\frac{2Z_a}{r_{1a}}\right) u_i(1) + \sum(j) n_j \int u_j^*(2) u_j(2) \left(\frac{2}{r_{12}}\right) dv_2 u_i(1)$$
$$- 6\alpha \left[\frac{-3\rho\uparrow(1)}{4\pi}\right]^{1/3} u_i(1) = \varepsilon_i u_i(1) \quad (2\text{-}4)$$

provided we are dealing with a spin orbital of spin up. This is of the form of Equations 1-2 and 1-3, with

$$V_{X\alpha}(1) = 6\alpha \left[\frac{-3\rho\uparrow(1)}{4\pi}\right]^{1/3} \quad (2\text{-}5)$$

The value of the last term of Equation 2-5 corresponding to the free-electron gas is again given by $\alpha = 2/3$. The value originally suggested by the author was $\alpha = 1$. As we see next, a value between these two numbers proves to be nearer the truth.

To get reasonable values for α, K. Schwarz has made solutions of Equation 2-4 for many of the atoms and has recommended values of α given, running from about 0.75 for light atoms to about 0.69 for heavy atoms. He has shown that these α's have the following properties. First, if we use the spin orbitals arising from the solutions of Equation 2-4 and use these spin orbitals to compute the various terms of the Hartree–Fock total energy of an atom by Equation 1-6, we find that the ratio of the potential to the kinetic energy as computed from this equation is exactly -2, as is demanded by the virial theorem. Second, as shown by Schwarz and J. W. D. Connolly, the total Hartree–Fock energy of Equation 1-6 for an atom computed from these $X\alpha$ spin orbitals proves to be only very slightly higher than the energy computed from the best Hartree–Fock solutions for the same atom. Thus it appears that the $X\alpha$ method is capable of giving very good approximations to the Hartree–Fock wave functions and energy. The accuracy of the $X\alpha$ atomic orbitals is about the same as that of the so-called double-zeta orbitals of Clementi, which prove to be accurate enough for the type of LCAO calculations described in Chapter 1.

In the remainder of this book, we use Equation 2-3 for the total energy of a molecular system, and Equation 2-4 as the one-electron Schrödinger equation we are trying to solve. We describe many properties of this approximation later, but first we devote some time to practical methods of solving this $X\alpha$ problem for arbitrary molecular systems and atomic clusters.

3 Schrödinger's equation for the spherical problem

For the next several chapters, we consider individual atoms, later going to the molecular case.

Schrödinger's equation for the motion of an electron in the potential V, as given in Equation 1-3, can be written as

$$\nabla^2 u_i = -(\varepsilon_i + V)u_i \qquad (3\text{-}1)$$

where V in the self-consistent field method is given in Equation 1-2 and in the $X\alpha$ method by Equation 2-4. The $X\alpha$ method is simpler than the general self-consistent field method in that V is the same for all electrons, so that we do not have to write it as V_i in Equation 1-2. In the present chapter we take up the case in which V has spherical symmetry. This is the case found in an isolated atom, and also in many of the molecular applications we make.

In the case of spherical symmetry we can make a separation of variables in spherical coordinates, writing u_i as the product of a function R_i of the radius vector r and a function $Y_{lm}(\theta, \phi)$, where θ and ϕ are the angles in spherical coordinates and l and m are quantum numbers, familiar from the case of the hydrogen atom. This separation arises from the form of the Laplacian ∇^2 in spherical coordinates. We have

$$\nabla^2 u = \frac{1}{r^2}\frac{\partial}{\partial r}\left(r^2 \frac{\partial u}{\partial r}\right) + \frac{1}{r^2 \sin\theta}\frac{\partial}{\partial \theta}\left(\sin\theta \frac{\partial u}{\partial \theta}\right) + \frac{1}{r^2 \sin^2\theta}\frac{\partial^2 u}{\partial \phi^2} \qquad (3\text{-}2)$$

If we substitute $R_i Y_{lm}(\theta,\phi)$ for u_i in Equation 3-1, multiply by r^2, and divide by $R_i Y_{lm}$, we have

$$\frac{1}{R_i}\left[\frac{d}{dr}\left(r^2 \frac{dR_i}{dr}\right) + (\varepsilon_i + V)r^2 R_i\right]$$
$$= -\frac{1}{Y_{lm}}\left[\frac{1}{\sin\theta}\frac{\partial}{\partial \theta}\left(\sin\theta \frac{\partial Y_{lm}}{\partial \theta}\right) + \frac{1}{\sin^2\theta}\frac{\partial^2 Y_{lm}}{\partial \phi^2}\right] \qquad (3\text{-}3)$$

On the left we have a function of r only, on the right a function of θ and ϕ only. Each of these must then be a constant. It is well known that the spherical harmonics are functions of θ and ϕ that satisfy the equation

$$\frac{1}{\sin\theta}\frac{\partial}{\partial\theta}\left(\sin\theta\,\frac{\partial Y_{lm}}{\partial\theta}\right) + \frac{1}{\sin^2\theta}\frac{\partial^2 Y_{lm}}{\partial\phi^2} + l(l+1)Y_{lm} = 0 \qquad (3\text{-}4)$$

where l and m are integers. Thus if we write the constant value of each side of Equation 3-3 in the form $l(l+1)$, we see that Y_{lm} will be a spherical harmonic, whose properties we go into in detail later. The function R_i will then satisfy the equation

$$\frac{1}{r^2}\frac{d}{dr}\left(r^2\frac{dR_i}{dr}\right) + \left[\varepsilon_i + V - \frac{l(l+1)}{r^2}\right]R_i = 0 \qquad (3\text{-}5)$$

We consider this Equation 3-5 in the present chapter.

The first step in the treatment of Equation 3-5 is to replace R_i by a function P_i/r, so that $P_i = rR_i$. Then Equation 3-5 is converted into

$$\frac{d^2 P_i}{dr^2} = g(r)P_i(r) \qquad (3\text{-}6)$$

where

$$g(r) = -\left[\varepsilon_i + V - \frac{l(l+1)}{r^2}\right] \qquad (3\text{-}7)$$

There are, of course, several special cases in which Equation 3-6 can be solved exactly. First is that of the hydrogen atom, in which V is $2/r$. The solutions are so well known that we do not discuss them. Second, there is the case in which V is zero. In this case the solutions are spherical bessel or neumann functions, which we discuss later. For both the hydrogen case and the case of zero V the nature of the solution is quite different, depending on whether the energy ε_i is positive or negative. For the hydrogen case the negative energies yield solutions for the stationary states, whereas positive energies give a continuum. Apart from these special cases, in general there is no closed analytic method for solving Equation 3-6 for the type of $g(r)$ which we meet in an atom.

The workers in the field, from Hartree on, have concluded that the most practical method for handling Equation 3-6 is numerical integration, constructing a table of values of the function. There is a very simple but rather inaccurate way of constructing such a table of values. Suppose we have found values of the function P at a set of equal spaced values of r, namely, ... P_{n-2}, P_{n-1}, P_n, differing by increments h of the variable r. Let us suppose we know the expansion of P in power series around the value r_n of r involved in

3 SCHRÖDINGER'S EQUATION FOR THE SPHERICAL PROBLEM

P_n. We then have, from the Taylor expansion,

$$P_{n\pm 1} = P_n \pm hP'_n + \frac{h^2}{2}P''_n \pm \frac{h^3}{6}P'''_n + \frac{h^4}{24}P^{iv}_n \pm \frac{h^5}{120}P^{v}_n + \frac{h^6}{720}P^{vi}_n \cdots \quad (3\text{-}8)$$

where P'_n, P''_n, and so forth are the successive derivatives of P with respect to r, computed at r_n. Then we note that

$$P_{n+1} - 2P_n + P_{n-1} = h^2 P''_n + \frac{h^4}{12}P^{iv}_n + \cdots \quad (3\text{-}9)$$

or

$$P_{n+1} = (2 + g_n h^2)P_n - P_{n-1} + \frac{h^4}{12}P^{iv}_n + \cdots \quad (3\text{-}10)$$

If the intervals h are small enough so that the fourth order term $h^4 P^{iv}_n/12$ can be neglected, this allows us to compute the next entry P_{n+1} in the table if P_n, P_{n-1} are known, and by repetition to get the whole table.

Generally this approximation is not accurate enough. In this case it is customary to use a modification of the procedure suggested by Noumerov, which is more accurate. We define a quantity y_n by the definition

$$y_n = P_n - \frac{h^2 P''_n}{12} \quad (3\text{-}11)$$

Then we can verify the relation, similar to Equation 3-9,

$$y_{n\pm 1} = P_n \pm hP'_n + \frac{5}{12}h^2 P''_n \pm \frac{h^3}{12}P'''_n \mp \frac{h^5}{180}P^{v}_n - \frac{h^6}{480}P^{vi}_n \cdots \quad (3\text{-}12)$$

from which we find

$$y_{n+1} - 2y_n + y_{n-1} = h^2 P''_n - \frac{h^6}{240}P^{vi}_n \cdots \quad (3\text{-}13)$$

showing that the fourth-order term has disappeared. We may rewrite Equation 3-11 by use of Equation 3-6, and we have

$$y_n = P_n\left(1 - \frac{g_n h^2}{12}\right) \quad (3\text{-}14)$$

from which

$$y_{n+1} = \left(2 + \frac{g_n h^2}{1 - g_n h^2/12}\right)y_n - y_{n-1} - \frac{h^6}{240}P^{vi}_n \cdots \quad (3\text{-}15)$$

Here the sixth-order term is small enough to neglect in the actual cases we encounter, and Equation 3-15 is almost as convenient to compute as Equation 3-10, and much more accurate. To carry out a Noumerov integration in a practical way we first construct a table of values of the quantities $g_n h^2/12$, which can be done from Equation 3-7. Then we carry through two columns, one of y_n, the other of P_n. From the entries in the y_n table, up to y_n itself, we can use Equation 3-15 (neglecting the term in h^6) to find y_{n+1}, and from Equation 3-14 we get P_{n+1}.

We now wish to give a practical example of Noumerov integration, so that the reader can see for himself how it works. Herman and Skillman have used this method to compute orbitals in the ground state of all the atoms, carrying their calculations to self-consistency. There are several comments concerning their results, which are tabulated in their book. First, they are carried out with $\alpha = 1$, since they were done before the days in which different values of α were considered. Second, they used what is called the Latter correction. Latter had suggested that rather than using the $X\alpha$ type of potential out to infinite r, one should join the potential V to a limiting value of $2/r$ at large r. Later work has shown that this was a poor idea, and present calculations are being made without using this correction, but one must realize that the Herman–Skillman tabulated calculations for large r are slightly modified by its use. The modification, which affects only the exact shape of the tail of the wave function at large r, is hardly noticeable, but it must be taken into account.

In addition to these two points, one must be familiar with the so-called Herman–Skillman mesh of points at which the function is computed. Herman and Skillman have used as their independent variable not r, measured in bohrs, but a quantity x, equal to r/μ, where $\mu = (9\pi^2/128Z)^{1/3}$, Z being the atomic number. Instead of tabulating the potential V of Equation 3-7, they use the quantity

$$U = \frac{\mu x}{2Z} V(x) \qquad (3\text{-}16)$$

The reason for this choice is that U, as a function of x, follows very similar curves for all atoms of the periodic table, going from unity when $x = 0$ down to zero at large x. It is a function that, on the simplified Thomas–Fermi model of the atom, is identical for all atoms.

Herman and Skillman use for their published tables what they call a 110-point presentation mesh, consisting of 11 blocks of numbers. We indicate these by giving the initial and final x and the interval for each block: 0(0.01)0.1; 0.1(0.02)0.3; 0.3(0.04)0.7; 0.7(0.08)1.5; 1.5(0.16)3.1; 3.1(0.32)6.3; 6.3(0.64)12.7; 12.7(1.28)25.5; 25.5(2.56)51.1; 51.1(5.12)102.3; 102.3(10.24)204.7.

3 SCHRÖDINGER'S EQUATION FOR THE SPHERICAL PROBLEM

For actual calculations they use what they call an integration mesh, with points spaced one-fourth as far apart as in the presentation mesh. They give tabulations of U to five significant figures at each point of the presentation mesh, and also the various P_i's in the occupied orbitals of the atom to four significant figures.

To illustrate the Noumerov method, we now give in Table 3-1 the quantities $gh^2/12$, P_{2s}, and y_{2s} for the 2s orbital of the carbon atom, $Z = 6$, for which $\mu = 0.4872221379$, so that $x = 1$ corresponds to about a half a bohr unit. The quantities $g_n h^2/12$ are computed from Herman and Skillman's tabulated values of U, which are listed in Note 1. The value of ε_{2s} is -1.2895 rydbergs. We give two values of P_{2s}: the four-figure value tabulated by Herman and Skillman, and given in Note 1, and that derived from our Noumerov integration. We include the blocks going from $x = 0.3$ to 12.7, five out of the eleven Herman–Skillman blocks. Instead of using an integration mesh with 4 times as many points as the presentation mesh, we have used the presentation mesh itself for our Noumerov integration. The reader with a pocket calculator giving 10-place accuracy will be able to verify the correctness of Equation 3-15 (omitting the sixth-order term) when applied to the entries in the table. It is clear that the agreement between our Noumerov calculation and the Herman–Skillman values is satisfactory: except for the very last entries in the table, the discrepancies between our calculated values and those of Herman and Skillman are no greater than one unit in the last

Table 3-1 Noumerov integration for the 2s orbital of carbon, following Herman and Skillman

x	$g_n h^2/12$	P_{2s}(HS)	P_{2s}(Noumerov)	y_{2s}
0.26	$-2.437032422 \times 10^{-3}$	0.3597	0.3596759469	0.3605524887
0.30	$-2.044990003 \times 10^{-3}$	0.3595	0.3595036911	0.3602388726
0.34	$-1.748035944 \times 10^{-3}$	0.3505	0.350490409	0.3511030789
0.38	$-1.515978225 \times 10^{-3}$	0.3341	0.3341087456	0.3346152471
0.42	$-1.330089425 \times 10^{-3}$	0.3116	0.3116348943	0.3120493964
0.46	$-1.178156010 \times 10^{-3}$	0.2841	0.2841747161	0.2845095183
0.50	$-1.051872777 \times 10^{-3}$	0.2526	0.2526862206	0.2529520144
0.54	$-9.453953233 \times 10^{-4}$	0.2180	0.2179988903	0.2182049854
0.58	$-8.545452577 \times 10^{-4}$	0.1808	0.18083202872	0.1809848148
0.62	$-7.761973363 \times 10^{-4}$	0.1417	0.1417986639	0.1419087276
0.66	$-7.080311965 \times 10^{-4}$	0.1014	0.1014400527	0.1015118754
0.70	$-6.482457077 \times 10^{-4}$	0.0602	0.0602141169	0.0602531505

Table 3-1 (*continued*)

x	$g_n h^2/12$	P_{2s}(HS)	P_{2s}(Noumerov)	y_{2s}
0.62	$-3.104789345 \times 10^{-3}$	0.1417	0.1417086639	0.1422389188
0.70	$-2.592982831 \times 10^{-3}$	0.0602	0.0602141169	0.0603702510
0.78	$-2.194437549 \times 10^{-3}$	-0.0234	-0.0233208507	-0.0233720269
0.86	$-1.877291904 \times 10^{-3}$	-0.1064	-0.1063006337	-0.1065001910
0.94	$-1.620962046 \times 10^{-3}$	-0.1870	-0.1869306598	-0.1872336673
1.02	$-1.411413518 \times 10^{-3}$	-0.2640	-0.2639584989	-0.2643310536
1.10	$-1.238467828 \times 10^{-3}$	-0.3336	-0.3365409895	-0.3369577847
1.18	$-1.094495353 \times 10^{-3}$	-0.4042	-0.4041406437	-0.4045829736
1.26	$-9.734081963 \times 10^{-4}$	-0.4665	-0.4664461594	-0.4669002018
1.34	$-8.704093571 \times 10^{-4}$	-0.5234	-0.5233134232	-0.5237689199
1.42	$-7.818268207 \times 10^{-4}$	-0.5748	-0.5747223418	-0.5751716752
1.50	$-7.04751115 \times 10^{-4}$	-0.6208	-0.6207449595	-0.6211824301
1.34	$-3.481637429 \times 10^{-3}$	-0.5234	-0.5233134232	-0.5251354106
1.50	$-2.819004460 \times 10^{-3}$	-0.6208	-0.6207449595	-0.6224994906
1.66	$-2.308152087 \times 10^{-3}$	-0.6973	-0.6972554488	-0.6988648204
1.82	$-1.901374641 \times 10^{-3}$	-0.7545	-0.7544831361	-0.17559176914
1.98	$-1.570166126 \times 10^{-3}$	-0.7946	-0.7945083916	-0.7957559016
2.14	$-1.296061105 \times 10^{-3}$	-0.8196	-0.8195617875	-0.8206239896
2.30	$-1.066654259 \times 10^{-3}$	-0.8319	-0.8318582195	-0.8327455244
2.46	$-8.728690707 \times 10^{-4}$	-0.8335	-0.8334918682	-0.8342193974
2.62	$-7.078461921 \times 10^{-4}$	-0.8264	-0.8263779713	-0.8269629196
2.78	$-5.664400019 \times 10^{-4}$	-0.8122	-0.8122269818	-0.8126870596
2.94	$-4.445401657 \times 10^{-4}$	-0.7925	-0.7925042932	-0.7928565931
3.10	$-3.388446988 \times 10^{-4}$	-0.7685	-0.7685381111	-0.7687985264
2.78	$-2.265760007 \times 10^{-3}$	-0.8122	-0.8122269818	-0.8140672932
3.10	$-1.355378795 \times 10^{-3}$	-0.7685	-0.7685381111	-0.7695797715
3.42	$-6.646693402 \times 10^{-4}$	-0.7121	-0.7121190034	-0.7125923268
3.74	$-1.303085141 \times 10^{-4}$	-0.6498	-0.6498403185	-0.6499249985
4.06	$2.900797612 \times 10^{-4}$	-0.5863	-0.5864116204	-0.5862415142
4.38	$6.254625944 \times 10^{-4}$	-0.5248	-0.5249276261	-0.5245993030
4.70	$8.428865916 \times 10^{-4}$	-0.4672	-0.4672908371	-0.4668969638
5.02	$9.556667784 \times 10^{-4}$	-0.4143	-0.4143170515	-0.4139211024
5.34	$1.054930238 \times 10^{-3}$	-0.3661	-0.3660828210	-0.3656966293
5.66	$1.142969562 \times 10^{-3}$	-0.3225	-0.3224750371	-0.3221064581
5.98	$1.221586616 \times 10^{-3}$	-0.2833	-0.2832852942	-0.2829392366
6.30	$1.292217176 \times 10^{-3}$	-0.2483	-0.2482454925	-0.2479247055
5.66	$4.571878247 \times 10^{-3}$	-0.3225	-0.3224750371	-0.3210007205
6.30	$5.168868704 \times 10^{-3}$	-0.2483	-0.2482454925	-0.2469623441
6.94	$5.655751412 \times 10^{-3}$	-0.1895	-0.1893929073	-0.1883217481
7.58	$6.060416460 \times 10^{-3}$	-0.1436	-0.1434041515	-0.1425350626
8.22	$6.402067971 \times 10^{-3}$	-0.1081	-0.1078680222	-0.1071774437
8.86	$6.694361252 \times 10^{-3}$	-0.0810	-0.0806466434	-0.0801067657
9.50	$9.947271858 \times 10^{-3}$	-0.0604	-0.0599309776	-0.0595146209
10.14	$7.168256866 \times 10^{-3}$	-0.0449	-0.0442358515	-0.0439187576
10.78	$7.363002466 \times 10^{-3}$	-0.0332	-0.0323663353	-0.0321280219
11.42	$7.535920188 \times 10^{-3}$	-0.0245	-0.0233731855	-0.0231970470
12.06	$7.690485117 \times 10^{-3}$	-0.0180	-0.0165066779	-0.0163797336
12.70	$7.829471848 \times 10^{-3}$	-0.0132	-0.0111732330	-0.0110857525

significant figure tabulated by Herman and Skillman. There are a number of observations to be made about this agreement.

In the first place, the reader should realize that this is a type of calculation in which errors can build up very rapidly. Each entry in the table depends on the preceding ones, and this makes it necessary to carry all 10 figures of which the calculator or computer is capable, in order that roundoff errors should not get too large. In the second place, the reader may well ask, how did we get the first two entries in the table, y_{2s} for $x = 0.26$ and 0.30, on which all other entries are based? This is a crucial question, and it is taken up in detail in the next chapter. To appreciate how essential it is to have the correct starting values, we may mention that if we had taken for the initial values of P_{2s} the values 0.3597, 0.3595 from the Herman–Skillman calculation, rather than the 10-figure entries in the table, errors would have built up fast enough so that the calculated function for $x = 0.70$ would have been about 0.0600 rather than 0.0602. This error would have increased in the subsequent blocks of the calculation and would have given a substantial error in the last block we have tabulated.

In spite of the accuracy of our calculation, the reader will notice that our calculated values of P_{2s} are becoming smaller in absolute value than the tabulated Herman–Skillman values as we approach the last entry in the table, that for 12.70. This type of behavior is the sign of a slightly incorrect value of the eigenvalue. We have used the value -1.2895 for the energy parameter (in rydbergs) listed by Herman and Skillman. The P_{2s} we have found actually goes through zero at a value of x of about 14.62, whereas the correct function is falling down asymptotically to zero. If a slightly different energy parameter had been assumed, the function would not have changed sign at large x, but instead would have gone through a minimum of absolute value and then would have started to rise again. Some workers determine the correct eigenvalue by interpolation between the behavior of the functions at slightly different energies, in the way described. It should be said that to get as good agreement as we have obtained between the result of a Noumerov integration out to very large values of x, and a correct function, indicates very accurate work and a very close approach to the correct eigenvalue. We shall come back to this point concerning the behavior of the function at large x in later chapters, particularly Chapter 12. We prefer not to make a point of it here, since we have already mentioned that on account of the use of the Latter correction, the Herman–Skillman wave functions are not very close approximations to the correct values in this region.

4 Power series expansions in the spherical problem

In the preceding chapter we asked the question, how are the two initial values of y_{2s} in a Noumerov expansion to be determined? We now answer this question. Starting with Hartree, everyone who has handled this problem has realized that a power series expansion of the potential and the wave function must be made to get a solution of Schrödinger's equation for the atom holding near the nucleus. The nuclear potential, going as $1/r$, leads to a singularity in Schrödinger's equation, but a convergent power series for P is possible about $r = 0$, just as the solutions of the hydrogen problem can be written as the product of an exponential $\exp(-ar)$, which in turn can be expanded in power series and of a polynomial in r. The values of the function given by this power series expansion can be joined onto a Noumerov integration at some suitable value of r.

Hartree made a very rudimentary expansion of the potential, fitting the power series for the potential to the tabulated values at several values of x near zero, and then he substituted this expansion into Schrödinger's equation, from which he could determine the coefficients of the first few terms in the expansion of the quantity P. He used this method to determine the first entries in the table (which would correspond, in terms of the Herman–Skillman integration mesh, to $x = 0.0025, 0.005$) and used Noumerov integration from that point on. Subsequent workers have followed this same procedure. This, however, does not really take advantage of the capabilities of the power-series method.

In the present chapter we show that, in fact, one can get a series expansion which is so accurate that it can be used up to about $x = 0.5$, making possible very accurate starting values for the quantities y_{2s} at $x = 0.26, 0.30$, where we started the Noumerov integration in Table 3-1. Thus it is possible to eliminate the need for Noumerov integration in the first two blocks of the Herman–Skillman mesh. Furthermore, the cumulative errors arising from these two initial blocks are eliminated, making it possible to use the Noumerov integration with the presentation mesh used in Table 3-1 instead of the fourfold finer integration mesh of Herman and Skillman. Alternatively, by using the series to compute values of the function at $x = 0.01, 0.02$, one can start the Noumerov integration at $x = 0$, using the presentation mesh. By these means we greatly reduce the computational labor of inte-

4 POWER SERIES EXPANSIONS IN THE SPHERICAL PROBLEM

grating Schrödinger's equation. Almost all present workers use computer programs for this integration, although the original Hartree calculations were made with a desk calculator. But with the reduction in labor made by the methods we are proposing, it is entirely practical to carry through an integration with a pocket calculator (as the calculation for Table 3-1 was done), although, of course, the time- and labor-saving advantages of a computer program should be used whenever large calculations are to be made.

The first thing to do, as Hartree did, is to get a power series expansion of the quantity $g(r)$ of Equations 3-6 and 3-7 in powers of r. There are two terms we know exactly: the quantity $l(l+1)/r^2$ and the nuclear potential term $-2Z/r$. The energy ε_i is, of course, a constant term, and this leaves only the two quantities V_e and V_X of Equation 1-2, that are not simple analytic functions. Our problem is to see if the sum of these two terms can be expressed as a power series. As a first step we should find out just what sort of functions we are dealing with. We postpone to the next chapter the problem of determining these quantities, from Equation 2-4, in terms of the actual electronic charge density. Rather, we now take advantage of the fact that those who have solved a self-consistent field problem have had to go through these steps, and we use Herman and Skillman's tabulation of the quantity U of Equation 3-16. We use their tabulation of U for carbon, which we have already used without much explanation in determining the quantity $g_n h^2/12$ in Table 3-1, and which is given in Note 1, as an example.

In Table 4-1 we give values of $-V_e - V_X$ as a function of x for carbon, determined from Herman and Skillman's values of U by the equation

$$-V_e - V_X = \frac{2Z}{r}(1 - U) \tag{4-1}$$

This equation simply says that the whole potential, which by Equation 3-16 is $(2Z/r)U$ in rydbergs, is the sum of the nuclear potential $2Z/r$ and the electronic potentials $V_e + V_X$. We carry this table only out to $x = 0.70$, since that is as far as we need it for present purposes; further information is given in Table 4-4. For comparison, we also give the nuclear potential $2Z/r$. The striking feature of Table 4-1 is the smooth nature of the electronic potential $-V_e - V_X$ as a function of x. It gradually decreases to zero as x becomes infinite, but it is obviously a function that can be expanded perfectly well in power series.

Table 4-1 Nuclear potential $2Z/r = 12/\mu x$ (in rydbergs), and electronic potential $-V_e - V_x$ of Equation 4-1, as function of x, for x up to 0.70 for carbon. Last column, electronic potential as determined from polynomial expansion, to be described later

x	$2Z/r$	Electronic potential $-V_e - V_x$ (HS)	Electronic potential $-V_e - V_x$ (polynomial)
0.01	2462.942273	15.171724	15.19095013
0.02	1231.471137	15.418019	15.43550280
0.03	820.9807577	15.6396834	15.64687005
0.04	615.7355683	15.8367188	15.83852766
0.05	492.5884546	16.0091248	16.00805394
0.06	410.4903788	16.1569013*	16.15690148
0.07	351.8488961	16.2870854	16.28644810
0.08	307.8677841	16.3970381	16.39800011
0.09	273.6602526	16.4907668	16.49279502
0.10	246.2942273	16.5682127	16.57200438
0.12	205.2451894	16.6772766	16.68804004
0.14	175.9244481	16.7427297	16.75426483
0.16	153.9338921	16.7618615	16.77795168
0.18	136.8301263	16.7480075	16.76556231
0.20	123.1471137	16.7036745	16.72282054
0.22	111.9519215	16.6349360	16.65477750
0.24	102.6225947	16.54686716	16.56587330
0.26	94.72854896	16.44298154	16.45999405
0.28	87.96222404	16.32578879	16.34052536
0.30	82.09807577	16.19877134	16.21040170
0.34	72.43947862	15.92219741	15.92793180
0.38	64.81427034	15.62866495*	15.62866499
0.42	58.64148269	15.32888357	15.32401554
0.46	53.54222333	15.02983750	15.02173879
0.50	49.25884546	14.73627620	14.72677050
0.54	45.61004209	14.45154183	14.44192200
0.58	42.46452195	14.17635600	14.16843417
0.62	39.72487537	13.91204860	13.90651977
0.66	44.78076860	13.65813443	13.65556719
0.70	35.18488961	13.41459101*	13.41459081

* Entries marked with an asterisk are used for fitting the polynomial to the Herman–Skillman value.

4 POWER SERIES EXPANSIONS IN THE SPHERICAL PROBLEM

The method we have found useful for describing such a function as $-V_e - V_x$ of Table 4-1 is expansion as a tenth order polynomial in x. The function $-V_e - V_x$ labeled electronic potential (polynomial) in the table is the following polynomial:

$$14.92339239 + 28.16365922x - 143.7359241x^2 + 304.8950032x^3$$
$$- 381.7567425x^4 + 303.8835406x^5 - 158.0733923x^6 + 53.64405392x^7$$
$$- 11.46126427x^8 + 1.400836216x^9 - 0.0747273802x^{10} \qquad (4\text{-}2)$$

This polynomial was obtained, using the method described in Note 2, by fitting exactly to the following Herman–Skillman entries for U:

$$x = 0.06, \ 0.96064; \qquad x = 0.38, \ 0.75887; \qquad x = 0.70, \ 0.61874;$$
$$x = 1.02, \ 0.51509; \qquad x = 1.34, \ 0.44420; \qquad x = 1.66, \ 0.39410;$$
$$x = 1.98, \ 0.35292; \qquad x = 2.30, \ 0.31711; \qquad x = 2.62, \ 0.28586;$$
$$x = 2.94, \ 0.25871; \qquad x = 3.26, \ 0.22992* \qquad (4\text{-}3)$$

One will notice from Table 4-1 that the polynomial and the Herman–Skillman values agree at $x = 0.06, 0.38$, and 0.70 with errors of a unit in the eighth significant figure. The discrepancies in the ninth and tenth figures arise from the round-off errors in deriving the polynomial from the numbers of Equation 4-3 and in the computation of the value of the polynomial of Equation 4-2.

Between these points at which the polynomial is intended to reproduce the Herman–Skillman values exactly, marked by asterisks in the table, there are very slight discrepancies between the Herman–Skillman values and the polynomial, which tend to alternate in sign from one interval to the next. Thus in the range of x from 0.06 to 0.38, the polynomial is as much as 0.02 rydbergs larger than the Herman–Skillman value, and from 0.38 to 0.70 it is as much as 0.01 rydbergs smaller. This type of alternation persists, with an amplitude that decreases with increasing x. Such oscillatory errors in the potential tend to neutralize each other in solving Schrödinger's equation, and if one is satisfied with a local error of not over 0.02 in the potential, one can use Equation 4-2 for the potential throughout the range from $x = 0.06$ (or for that matter, $x = 0$) up to $x = 2.94$.

We should point out what an economy in storage arises from the availability of this polynomial expansion, provided the accuracy just described is adequate. The numbers given in Equation 4-3 are all one requires to be able

* Interpolated

to reproduce a table of potentials, or of the quantity $g_n h^2/12$ of Table 3-1, all the way from $x = 0$ to $x = 2.94$. We shall see later that one or two further power series expansions allow us to carry this process through to the largest values of x we require. We find it profitable to use the power series solution of Schrödinger's equation, which we come to shortly, only about to $x = 0.3$, as we have done in Table 3-1. But from the polynomial expansion of the potential we have the necessary information to carry through a Noumerov integration over the complete range of the variable. Connolly has set up computer programs to carry out this process, verifying the statement that there is great economy in computer storage arising in this way. The wave functions we get in this way are in good agreement with the Herman–Skillman values, and the eigenvalues agree with the Herman–Skillman values with errors of the order of one part in 10^4.

We now see that we can write the quantity g of Equation 3-6 in the form of $l(l+1)/r^2 - 2Z/r$ plus a power series or polynomial in r. Let us write this polynomial in the form

$$g = \frac{l(l+1)}{r^2} - \frac{2Z}{r} + \sum (m) v_m r^m$$

$$= \frac{l(l+1)}{\mu^2 x^2} - \frac{2Z}{\mu x} + \sum (m) v_m \mu^m x^m \tag{4-4}$$

We express P as a power series,

$$P = \sum (n) p_n r^n = \sum (n) p_n \mu^n x^n \tag{4-5}$$

Let us then express Equation 3-6 in terms of these series expansions. We have

$$\frac{d^2 P}{dr^2} = \sum (n) n(n-1) p_n r^{n-2} \tag{4-6}$$

We can express the product gP in the form

$$gP = \sum (n,m) v_m p_{n-m-2} r^{n-2} \tag{4-7}$$

if we define $v_{-2} = l(l+1)$, $v_{-1} = -2Z$, $v_m = 0$ for $m < -2$. We equate the coefficients of equal powers of r in Equations 4-6 and 4-7, and obtain the recurrence relations

$$n(n-1) p_n = \sum (m) v_m p_{n-m-2} \tag{4-8}$$

From these recurrence relations, we can find values of all p_n's, once we have started the recurrence, and hence can get an explicit expression for the power series expansion of the function P. Let us write down these recurrence equations explicitly.

4 POWER SERIES EXPANSIONS IN THE SPHERICAL PROBLEM

We first rewrite Equation 4-8 in the form giving explicit values to v_{-2}, v_{-1}. We have

$$n(n-1)p_n = l(l+1)p_n - 2Zp_{n-1} + v_0 p_{n-2} + v_1 p_{n-3} \cdots \quad (4\text{-}9)$$

Obviously, we can combine the terms in p_n to give

$$[n(n-1) - l(l+1)]p_n = -2Zp_{n-1} + v_0 p_{n-2} + v_1 p_{n-3} \cdots \quad (4\text{-}10)$$

We expect from our experience with the hydrogenic problem that the first nonvanishing term in the expansion of P will have $n = l + 1$. For this term, the coefficient $n(n-1) - l(l+1)$ is zero, so that Equation 4-10 is identically satisfied with an arbitrary value of p_{l+1}, combined with $p_n = 0$ for $n < l + 1$. The quantity p_{l+1} is the arbitrary constant that will eventually be determined to normalize the wave function. We can then write the successive equations of Equation 4-10 in the following form:

$$(2l+2)p_{l+2} = -2Zp_{l+1}$$
$$2(2l+3)p_{l+3} = -2Zp_{l+2} + v_0 p_{l+1}$$
$$3(2l+4)p_{l+4} = -2Zp_{l+3} + v_0 p_{l+2} + v_1 p_{l+1} \quad (4\text{-}11)$$
$$4(2l+5)p_{l+5} = -2Zp_{l+4} + v_0 p_{l+3} + v_1 p_{l+2} + v_2 p_{l+1}$$
$$\cdots\cdots\cdots$$

Since we are dealing with a polynomial whose last term is v_{10}, we see that no one of these equations will have more than 12 terms on the right side, but we have an infinite number of equations, so that the wave function P is given by an infinite series in r.

Ordinarily, our calculations are made in terms of x rather than of r. Consequently, it is more convenient to follow the lead of Equation 4-5 and use as the coefficients the quantities $p_n \mu^n$. In terms of these quantities Equations 4-9 and 4-11 are rewritten in the form

$$[n(n-1) - l(l+1)]p_n \mu^n = \sum(m)v_m \mu^{m+2} p_{n-m-2} \mu^{n-m-2}$$
$$(2l+2)p_{l+2}\mu^{l+2} = (-2Z\mu)p_{l+1}\mu^{l+1}$$
$$2(2l+3)p_{l+3}\mu^{l+3} = (-2Z\mu)p_{l+2}\mu^{l+2} + v_0\mu^2 p_{l+1}\mu^{l+1}$$
$$3(2l+4)p_{l+4}\mu^{l+4} = (-2Z\mu)p_{l+3}\mu^{l+3} + v_0\mu^2 p_{l+2}\mu^{l+2} \quad (4\text{-}12)$$
$$\qquad\qquad + v_1\mu^3 p_{l+1}\mu^{l+1}$$
$$\cdots\cdots\cdots$$

It is consequently convenient to express the v_m's in the form $v_m \mu^{m+2}$.

We are now ready to apply this procedure, and we choose the example we have already been discussing, the 2s state of carbon. In Table 4-2 we tabulate

Table 4-2 Quantities $v_m\mu^{m+2}$, $p_n\mu^n$ for integration of Schrödinger's equation for $2s$ orbital of carbon

$v_{-1}\mu = -5.846665654$	$p_1\mu = 1.00000000$
$v_0\mu^2 = 3.848704135$	$p_2\mu^2 = -2.923332827$
$v_1\mu^3 = 6.685641839$	$p_3\mu^3 = 3.490075628$
$v_2\mu^4 = -34.12081152$	$p_4\mu^4 = -2.080892217$
$v_3\mu^5 = 72.37762586$	$p_5\mu^5 = -1.403330912$
$v_4\mu^6 = -90.62348149$	$p_6\mu^6 = 6.521785937$
$v_5\mu^7 = 72.13751939$	$p_7\mu^7 = -11.39846171$
$v_6\mu^8 = -37.52431731$	$p_8\mu^8 = 13.26833015$
$v_7\mu^9 = 12.73431583$	$p_9\mu^9 = -10.35081696$
$v_8\mu^{10} = -2.720736937$	$p_{10}\mu^{10} = 3.045031029$
$v_9\mu^{11} = 0.3325380819$	$p_{11}\mu^{11} = 6.34706536$
$v_{10}\mu^{12} = -0.0177391898$	$p_{12}\mu^{12} = -14.64967582$
	$p_{13}\mu^{13} = 18.99394122$
	$p_{14}\mu^{14} = -17.90366367$
	$p_{15}\mu^{15} = 11.86984295$
	$p_{16}\mu^{16} = -3.066324186$
	$p_{17}\mu^{17} = -5.622939674$
	$p_{18}\mu^{18} = 11.71713289$
	$p_{19}\mu^{19} = -13.90697628$
	$p_{20}\mu^{20} = 12.27315627$
	$p_{21}\mu^{21} = -7.999027315$
	$p_{22}\mu^{22} = 2.779236715$
	$p_{23}\mu^{23} = 1.810093583$
	$p_{24}\mu^{24} = -4.754815547$
	$p_{25}\mu^{25} = 5.755816928$
	$p_{26}\mu^{26} = -5.119624924$
	$p_{27}\mu^{27} = 3.497252698$
	$p_{28}\mu^{28} = -1.598740525$
	$p_{29}\mu^{29} = -0.0212132498$
	$p_{30}\mu^{30} = 1.064055709$
	$p_{31}\mu^{31} = -1.481633544$
	$p_{32}\mu^{32} = 1.401703659$
	$p_{33}\mu^{33} = -1.03069543$
	$p_{34}\mu^{34} = 0.5701746945$
	$p_{35}\mu^{35} = -0.1567205065$

the quantities involved in this power series integration of Schrödinger's equation. The quantity $v_{-1}\mu$ is -12μ, where we recall that for carbon $\mu = 0.4872221379$. The next entry, $v_0\mu^2$, is $(14.92339239 + 1.2895)\mu^2$, where $14.92339239 + 1.2895$ is the sum of the first term in Equation 4-2, the constant term in $-V_e - V_x$, and 1.2895, which is $-\varepsilon$ for the 2s orbital. The remaining terms, such as $v_1\mu^3$, are μ^2 times the coefficients of powers of x in Equation 4-2. In the listing of quantities $p_n\mu^n$, we have arbitrarily set the first entry equal to unity; later it can be multiplied by a constant required for normalization, which will multiply the whole function P.

The reader can then readily verify that Equations 4-12 are satisfied. We recall that for an s state, $l = 0$. Consequently, the first equation is $2p_2\mu^2 = v_{-1}\mu p_1\mu$, or $2(-2.923332827) = (-5.846665643)$. The second is $6(3.490075628) = (-5.846665654)(-2.923332827) + 3.848704135$, and so on. The entries in Table 4-2 can all be computed with a pocket calculator, but Connolly has programmed the calculation for a computer, with which it is quite practical to carry the series up to hundreds of terms. When we examine the convergence of the series, however, we find that it diverges if x is too large. It happens that the particular case shown in Table 4-2 is a convenient one to explain this fact, for we notice that all the coefficients $p_n\mu^n$ are of the order of magnitude of unity. Thus the convergence of the series will come only from the decrease in the quantity x^n as n increases, which means that there will be divergence if $x \geq 1$. If we demand that we use the 35-term series until the last term is of the order of 10^{-10}, to get ten-figure accuracy, we must require that $x^{35} = 10^{-10}$, $x = 0.518$. For values of x larger than this, the accuracy of the 35-term sum rapidly decreases, and even a great increase in the number of terms is unsuccessful in getting much more accuracy. We can safely use the series out to $x = 0.5$, but to be on the safe side we shall use it only to 0.3.

Let us then see how good a description of the function we can obtain with the series. In Table 4-3 we give values of the function P_{2s} calculated from the series, and we compare them with those given in the Herman–Skillman table, and where available, with the Noumerov solution of Table 3-1. In deriving the series values of P_{2s}, it is to be recalled that we must multiply the series by a constant. This constant was determined by computing the series with $p_1\mu = 1$ for all entries in the presentation mesh and dividing each entry by the corresponding Herman–Skillman value. To better than one part in a thousand, these ratios were all equal to 0.3166. Consequently, the series entries were multiplied by the arbitrary factor 1/0.3166 to derive the values of Table 4-3.

Table 4-3 Series values for the $2s$ orbital of carbon, compared with the Herman–Skillman and Noumerov entries of Table 3-1

x	P_{2s}(series)	P_{2s}(HS)	P_{2s}(Noumerov)
0.01	0.03067320	0.0307	
0.02	0.059561491	0.0596	
0.03	0.086738840	0.0867	
0.04	0.11225706	0.1123	
0.05	0.13617996	0.1362	
0.06	0.15856624	0.1586	
0.07	0.17947292	0.1795	
0.08	0.19895530	0.1990	
0.09	0.21706705	0.2171	
0.10	0.23386017	0.2339	
0.12	0.26369032	0.2637	
0.14	0.28882950	0.2888	
0.16	0.30963675	0.3096	
0.18	0.32644748	0.3265	
0.20	0.33957471	0.3396	
0.22	0.34931032	0.3493	
0.24	0.35592629	0.3559	
0.26	0.35967595	0.3597	0.35967595
0.28	0.36079521	0.3608	
0.30	0.35950369	0.3595	0.35950369
0.34	0.35049201	0.3505	0.35049041
0.38	0.33411268	0.3341	0.33410875
0.42	0.31164109	0.3116	0.31163489
0.46	0.28418417	0.2841	0.28417472
0.50	0.25269261	0.2526	0.25268622
0.54	0.21799511	0.2180	0.21799889
0.58	0.18077936	0.1808	0.18083203
0.62	0.14155519	0.1417	0.14179866
0.66	0.10046234	0.1014	0.10144005

In examining the values of Table 4-3, it is, first, clear that in practically every case the agreement between our series and the tabulated Herman-Skillman value is correct to the four-figure accuracy given by Herman and Skillman. Consequently, we conclude not only that in both cases Schrödinger's equation is being integrated correctly, but furthermore the slight discrepancies between the Herman–Skillman potential and the polynomial expansion, shown in Table 4-1, do not affect this four-place agreement between the two calculations. Second, we see where the entries for $x = 0.26$ and 0.30 came from, which were used to start the Noumerov expansion of Table 3-1. But furthermore, it is interesting to intercompare the series and Noumerov calculations in the range from $x = 0.30$ to 0.50, where both methods should be valid. We see that the discrepancies are less than one unit in the fifth significant figure, or one part in 30,000, throughout the range, giving us confidence in the accuracy of both methods. The agreement with the Herman–Skillman values is equally good by both methods, in no case leading to a discrepancy greater than one unit in the last digit of the Herman–Skillman value. Furthermore, it is interesting to find that the errors in the 35-term series values are not quite as serious as our earlier analysis would lead us to expect. It is not until $x = 0.62$ that the error in the series gets as large as two parts in a thousand, and the entry for $x = 0.66$ is the first one where the error in the series is as great as 1 percent.

We have thus far been discussing only a power series in r, or in x, expanding around the nucleus. It is natural to ask, could we not do something like the method of analytic continuation in mathematics, expanding around another point and hoping that there will be an overlapping region of convergence between the two expansions? The author has examined this question and has not been able to find an expansion that in all respects fulfills this hope. However, one can make very useful expansions in $x - x_0$, and we show how this is to be done.

In the first place, the methods of Note 2 show how an expansion of the potential as a tenth-order polynomial can be set up for this case, from the values of the function at 11 equally spaced points. One can set up a very useful polynomial by using the entries from the Herman–Skillman mesh corresponding to $x = 0.54, 1.18, 1.82, 2.46, 3.10, 3.74, 4.38, 5.02, 5.66, 6.30$, and 6.94. There is one fundamental difference between this expansion in $x - x_0$, where in this case we choose $x_0 = 3.74$, and the expansion in x. In the latter case, the nuclear potential in $1/x$ and the $l(l + 1)/r^2$ term in $1/x^2$ could be handled explicitly, requiring only an expansion of the quantity $-V_e - V_x$ of Table 4-1. In the case of the expansion in $x - x_0$ this is not possible, and we must expand the whole quantity g of Equation 3-7 in power series. Since this includes the terms in $1/x$ and $1/x^2$ which go infinite at the origin, it is obvious that the expansion of the potential will diverge when we approach the nucleus. It is found, in fact, that the series becomes inaccurate for a value of x considerably greater than the value $x = 0.5$ which we can reach with the

Table 4-4 V for carbon, as given by Herman and Skillman, as calculated from series in $x - x_0$ and from series in x

x	HS	Series in $x - x_0$	Series in x
0.42	43.31259912	40.62799206	43.31746717
0.46	38.51238583	37.14721391	38.52048448
0.50	34.52256926	34.00098128	34.53207498
0.54	31.15850026	*31.15850021	31.16812014
0.58	28.28816595	28.59198447	28.29607151
0.62	25.81282677	26.27568842	25.81835571
0.66	23.65917234	24.18600459	23.66174006
0.70	21.77029860	22.30128964	*21.77029887
0.78	18.62236968	19.06919582	18.61856336
0.86	16.11737967	16.44051995	16.11307862
0.94	14.09274607	14.29980504	14.09035777
1.02	12.43761701	12.55166269	*12.43761790
1.10	11.07159723	11.11780347	11.07328600
1.18	9.93442376	*9.93442275	9.93608460
1.26	8.97801100	8.94991627	8.97878041
1.34	8.16446984	8.12287343	*8.16447202
1.42	7.46479645	7.42035463	7.46424393
1.50	6.85601031	6.81639091	6.85554218
1.66	5.84726235	5.82760455	*5.84726544
1.82	5.04402458	*5.04402457	5.04414063
1.98	4.39000802	4.40053362	*4.39001234
2.14	3.84875012	3.86032777	3.8490549
2.30	3.39575489	3.40536544	*3.3957483
2.46	3.01309950	*3.01309950	3.0122585
2.62	2.68723923	2.68778142	*2.687221
2.78	2.40801335	2.40247335	2.411177
2.94	2.16730543	2.16373734	*2.167332
3.10	1.95859526	*1.95859526	1.933553
3.42	1.61762016	1.62208733	1.846
3.74	1.35382800	*1.35382800	
4.06	1.14629944	1.13955608	
4.38	0.98073462	*0.98073462	
4.70	0.87340126	0.87784957	
5.02	0.81772627	*0.81772627	
5.34	0.76872395	0.77438285	
5.66	0.72526252	*0.72526251	
5.98	0.68645249	0.67325435	
6.30	0.65158506	*0.65158501	
6.94	0.59149653	*0.59149632	

* Entries marked with an asterisk are used for fitting the polynomials to the Herman–Skillman values.

4 POWER SERIES EXPANSIONS IN THE SPHERICAL PROBLEM

expansion about the origin. It is for this reason that we must use the Noumerov method in the intermediate range.

By the methods of Note 2 we find for the potential $V = 2Z/r + V_e + V_x$ the following polynomial:

$$V = 1.353828001 - 0.7522078084(x - x_0) + 0.2567574446(x - x_0)^2$$
$$- 0.012906532(x - x_0)^3 + 0.0680795526(x - x_0)^4$$
$$- 0.0413181396(x - x_0)^5 - 0.0114042848(x - x_0)^6 + 7.148698278$$
$$\times 10^{-3}(x - x_0)^7 + 1.080150241 \times 10^{-3}(x - x_0)^8 - 6.580247853$$
$$\times 10^{-4}(x - x_0)^9 + 4.549423579 \times 10^{-5}(x - x_0)^{10} \quad (4\text{-}13)$$

where $x_0 = 3.74$. In Table 4-4 we give values of the quantity V as given by Herman and Skillman [equal to the quantity $2Z/r$ minus electronic potential (HS) from Table 4-1]. The second column is computed from Equation 4-13, and the third column is computed from the polynomial expansion in x given in Equation 4-2 and equal to the quantity $2Z/r$ minus electronic potential (polynomial) of Table 4-1.

We can make a number of comments about this table. First, we consider the series in x. The information given in this column, out to $x = 0.70$, is essentially already contained in Table 4-1, where we have noted the accuracy of this series over that range. It is interesting to see that this accuracy improves out to the limit of validity of this series, at $x = 2.94$. Beyond this point, we have included two additional entries, partly to show that the polynomial is rapidly departing from the correct function and partly to suggest that the accuracy of the calculation is rapidly being lost. The terms of the polynomial, in the limit of $x = 3.42$, are such very large numbers that even with the ten-place accuracy of the calculation, round-off errors reduce the accuracy of the result to four figures. Beyond this entry, it is quite impossible to compute the values of the function with the ten-place accuracy of the calculator. This difficulty is only academic, since we should not be using the polynomial in this range in any case.

Next we consider the series in $x - x_0$, which is the main point of the table. It is obvious that in the range of x up to 1.82, it is far less accurate than the series in x. Beyond this value of x, the discrepancies between the polynomial and the Herman–Skillman values are no larger than 0.02 rydbergs, comparable with the errors in the series in x for small values of x. We see evidence of the inaccuracy for x less than 1.82 later, when we consider the solution of Schrödinger's equation using the series in $x - x_0$ for the potential. In spite of this shortcoming, the series in $x - x_0$ is sufficiently accurate to be valuable over a very wide range, from $x = 1.82$ out to $x = 6.94$. We find that the resulting expansion of the wave function is of satisfactory accuracy over this same wide range. Hence there is only the range from $x = 0.5$ to $x = 1.82$ where we do not find adequate accuracy from the two power series, and where we must use the Noumerov integration.

Let us now consider the expansion of the wave function in the form

$$P = \sum(n) p_n \mu^n (x - x_0)^n \qquad (4\text{-}14)$$

which is analogous to Equation 4-5. We express the quantity g in the form

$$g = \sum(m) v_m \mu^m (x - x_0)^m \qquad (4\text{-}15)$$

which is exactly analogous to Equation 4-4, except that the terms $l(l+1)/\mu^2 x^2$ and $-2Z/\mu x$ are absent, being incorporated in the terms of the series with positive m. The recurrence relations are identical with those of Equation 4-12, if we set l and Z equal in those equations. In general, the coefficients p_n are nonvanishing for all positive values of n, but are necessarily zero for negative n.

There are two independent solutions of the problem, of which one can be chosen to have $p_0 = 1$, $p_1\mu = 0$, and the other has $p_0 = 0$, $p_1\mu = 1$. These two functions have an analogy to the cosine and sine, the first having zero slope, the second zero value at $x - x_0 = 0$. Consequently, we call them $C(x - x_0)$ and $S(x - x_0)$. The fact that we have two independent solutions is analogous to the fact that there are two independent solutions of our problem of expansion around the nucleus, one regular at the origin, the second diverging at the origin. We have been considering only the function regular at the origin, but in the expansion in $x - x_0$, we have use for both functions. We can give them arbitrary coefficients, thus setting up a solution with arbitrary function and slope at one arbitrary value of x.

From the potential V as given in Equation 4-13, we have the material for integrating the Schrödinger equation in a form as given in Table 4-2 for the two functions C and S. The reader should be able without trouble to construct such a table. We can immediately find the coefficient for the C function by matching the function to the exact function (either the Herman–Skillman or the Noumerov value) of Table 3-1 at $x = 3.74$. Then we can compute the difference between the function of Table 3-1 and the contribution of the C function, and we can see if this difference is proportional to the S function.

Table 4-5 Function P_{2s} of carbon, by Herman–Skillman, Noumerov integration (both from Table 3-1), and series in $x - x_0$

x	HS	Noumerov	Series in $x - x_0$
1.18	−0.4042	−0.4041406437	−0.4092973273
1.26	−0.4665	−0.4664461594	−0.4699501448
1.34	−0.5234	−0.5233134232	−0.5255955944
1.42	−0.5748	−0.5747223418	−0.5762002714
1.50	−0.6208	−0.6207449595	−0.6216865064
1.66	−0.6973	−0.6972554488	−0.6976642091
1.82	−0.7545	−0.7544831361	−0.7547325371
1.98	−0.7946	−0.7945083916	−0.7947650822
2.14	−0.8196	−0.8195617875	−0.8197996779
2.30	−0.8319	−0.8318582195	−0.832052213
2.46	−0.8335	−0.8334918682	−0.8337096357
2.62	−0.8264	−0.8263779713	−0.8265518277
2.78	−0.8122	−0.8122269818	−0.8121974002
2.94	−0.7925	−0.7925042932	−0.7924739046
3.10	−0.7685	−0.7685381111	−0.7685238677
3.42	−0.7121	−0.7121190034	−0.7121230615
3.74	−0.6498	−0.6498403185	−0.6498000000
4.06	−0.5863	−0.5864116204	−0.5863292776
4.38	−0.5248	−0.5249276261	−0.5248827962
4.70	−0.4672	−0.4672908371	−0.4672863923
5.02	−0.4143	−0.4143170515	−0.4143101546
5.34	−0.3661	−0.3660828210	−0.3660561071
5.66	−0.3225	−0.3224750371	−0.3222929735
5.98	−0.2833	−0.2832852942	−0.2823132944

This proves to be the case, over quite a wide range of values of $x - x_0$, and from this comparison we can get the coefficient of the S function. What we find is that the correct function is reproduced by the combination $-0.6498C + 0.4076S$. In Table 4-5 we give this function and for comparison the Herman–Skillman and Noumerov functions from Table 3-1.

As we examine the series in $x - x_0$, we note in the first place that it is quite poor for x below about 1.82. As we have already seen, the series expansion of the potential in this range is rather poor, and it is presumably for this reason that the wave function is not more accurate. On the other hand, from $x = 1.82$

up to 5.98, the series is quite acceptable, though not quite as close to the Herman–Skillman value as is the Noumerov integration. The main usefulness of the series in $x - x_0$ is as a very accurate interpolation formula for the wave function between the rather widely spaced values listed in the Herman–Skillman presentation mesh. The range of values of x included in the region of applicability of the series includes that part of the atom concerned in bond formation in molecules, and as we shall see later, we often need to know the wave function accurately at intermediate distances. The expansion in a power series is similar to a Lagrangian interpolation formula. In the conventional use of the Noumerov method, a separate routine must be set up for interpolating to get the values of the wave function at intermediate points.

We have seen the possibility of using power series expansions in $x - x_0$. It is very easy to set up additional power series with larger values of x_0, which give equally accurate descriptions of the wave function for larger x than 5.98. For instance, one can match the polynomial to the correct potential at $x = 5.02$, 5.66, 6.30, 6.94, 7.58, 8.22, 8.86, 9.50, 10.14, 10.78, 11.42, giving an expansion in $x - 8.22$. These entries are all contained in the Herman–Skillman presentation mesh. Where needed, still further ranges can be set up, overlapping those already used, and there is no reason to think that the wave functions would not be accurate. One can choose the coefficients of the functions C and S in each region so that the function matches that found from the preceding region in function and slope at an intermediate point at which both series are accurate. In other words, it is only the region between $x = 0.5$ and 1.82 that gives enough difficulty so that it must be bridged over by use of Noumerov integration. But even if the various power series expansions are to be used, it is very useful to have the Noumerov integrations as well, to evaluate the accuracy of the power series expansions.

We have been proceeding with the Herman–Skillman tables as a starting point. In practice, there would be no point in carrying this type of discussion through; we have done it for pedagogic reasons, to describe the various types of calculations that would be required. What one would actually meet would be a self-consistent field calculation, in which at a given stage of the iteration one would know a starting potential V, or the function U, and would need to find the various wave functions of the occupied orbitals, and hence to find a new potential to use in the next stage of iteration. We have presented enough calculational methods so that we can understand the solution of Schrödinger's equation. What we need now is to understand

5 Determination of the self-consistent field for atoms

In the preceding chapter we have studied the problem of integrating the Schrödinger equation to get atomic orbitals, provided we have the potential. Now we take up the other part of the problem, the determination of the potential if we know the orbitals. We shall continue to use the example of the Herman–Skillman calculation of the carbon atom, and again use the results tabulated in Note 1. The first problem we meet is the normalization of the orbitals. We recall that there is always an arbitrary constant, like the $p_1\mu$ of Table 4-2, which must be determined by normalization. When we write the atomic orbital in the form $P(r)Y_{lm}(\theta,\phi)$, the constants are such that if P satisfies the condition

$$\int_0^\infty P^2(r)\,dr = 1 \tag{5-1}$$

the square of the wave function will automatically integrate to unity over all space. We can immediately compute P^2 at all points of the presentation MESH from calculations given in Chapter 4 or from the Herman–Skillman values of Note 1. The problem is then simply to integrate a function given at a set of discrete points.

The standard way of doing this, having accuracy good enough for our purposes, is Simpson's rule. This rule simply passes a parabola through three neighboring points, P_{n-1}, P_n, and P_{n+1}, and computes the integral of the parabolic curve from point $n-1$ to point $n+1$. Its accuracy is equivalent to that of the simple method of Equations 3-8 to 3-10, disregarding fourth-power terms, but these neglected terms are less important in integration than in differentiation, so that a Simpson's rule integration using the presentation mesh is ordinarily accurate enough to use. We start with a Taylor's expansion of P,

$$P = P_n + (r - r_n)P'_n + \frac{(r - r_n)^2}{2} P''_n + \cdots \tag{5-2}$$

from which

$$\int_{r_n-h}^{r_n+h} P\, dr = 2h\left(P_n + \frac{h^2}{6} P_n'' + \cdots\right) \tag{5-3}$$

From Equation 3-9 we write $h^2 P_n'' = P_{n+1} - 2P_n + P_{n-1}$. Thus we find

$$\int_{r_n-h}^{r_n+h} P\, dr = \frac{h}{3}(P_{n-1} + 4P_n + P_{n+1}) \tag{5-4}$$

Table 5-1 Simpson integration of P_{1s}^2, for normalization integral of the carbon $1s$ orbital

x	P_{1s}^2	$\int_0^x P_{1s}^2\, dr$
0.00	0.00000000	0.00000000
0.01	0.01766241	
0.02	0.06666724	0.0002230127
0.03	0.14152644	
0.04	0.23726641	0.0016360209
0.05	0.34975396	
0.06	0.47499664	0.0050648937
0.07	0.60996100	
0.08	0.75168900	0.0110196083
0.09	0.89737729	
0.10	1.04530176	0.0197676817
0.12	1.34003776	
0.14	1.62435025	0.0458496670
0.16	1.88952516	
0.18	2.13043216	0.0825955807
0.20	2.34365481	
0.22	2.52778201	0.1281762947
0.24	2.68206129	
0.26	2.80696516	0.1803512645
0.28	2.90361600	
0.30	2.97424516	0.2368549821
0.34	3.04328025	
0.38	3.03108100	0.3549475406
0.42	2.95530281	
0.46	2.83181584	0.4698287785

5 DETERMINATION OF THE SELF-CONSISTENT FIELD FOR ATOMS

Equation 5-4 is a simple form of Simpson's rule. This rule can be used to find definite integrals or indefinite integrals up to a given entry in the table, but we note that there will be only half as many entries in the integral table as in the table of the original function. A second integration, using the intermediate entires of the table, can be made if we need as many entries as in the original function. To show just how it works out, in the cases we meet with the Herman–Skillman presentation mesh, we show in Table 5-1 the beginning of the integration of P_{1s}^2, met in the normalization of the 1s orbital and in the determination of how much of the 1s charge is located within a sphere of radius r_n. The P_n's are the Herman–Skillman values of Note 1.

In the first block of numbers, $h = 0.01\mu$, where we recall that for carbon $\mu = 0.4872221379$. Thus to get the integral to 0.02 we have

$$(0.01\mu/3)[4(0.01766241) + 0.06666724] = 2.230127461 \times 10^{-4},$$

to the accuracy of the calculator. The entry at 0.04 is $(0.01\mu/3)[0.06666724 + 4(0.14152644) + 0.23726641] + 0.0002230127 = 0.0016360209$, and so forth. It is trivial to note that if one wishes only the definite integral, one replaces Equation 5-4 by one in which the first and last entries have the coefficient $h/3$, the second and every alternating entry have $4h/3$, and the intermediate ones have $2h/3$. Furthermore, we note that we must have an odd number of entries in the sequence, as we have P_0, P_1, P_2, P_3, P_4 in determining the integral from zero to 0.04. In the second block of numbers h must be doubled, as it must be in going to each successive block. Part of the convenience of the Herman–Skillman mesh is that it is set up in such a way that Simpson's rule integration can be carried over so conveniently from one block to the next. Thus for finding the integral from 0.10 to 0.14 we simply use the three entries for 0.10, 0.12, and 0.14.

This shows us how to carry out the integration. It is, of course, assumed that the Ps in the Herman–Skillman tabulation are normalized, but one must expect a finite error in the integral from zero to infinity on account of the inherent error in Simpson's rule, plus the fact that the Ps are only given to four significant figures. It is found, in fact, that the integral for the 1s orbital, computed as in Table 5-1 by extending the table until the contributions become vanishingly small, is 1.000011619. Correspondingly for the 2s it is 1.000003424, and for the 2p it is 0.9999925445. The reader can get practice in Simpson's rule integration by computing P^2 for each of the orbitals from Note 1 and integrating to verify these numbers.

Table 5-2 Radial charge density, integral of charge, and actual charge density, for carbon atom, from Herman and Skillman

x	$\sum P^2$	$\int_0^x \sum P^2 \, dr$	$-\rho = \sum P^2 / 4\pi r^2$
0.02	0.14043952	0.00046981	117.6970167
0.04	0.49976790	0.00344604	104.7091141
0.06	1.00036170	0.01066779	93.1516742
0.08	1.58276050	0.02320728	82.90320853
0.10	2.20044244	0.04162494	73.76428247
0.14	3.41694786	0.09650984	58.44110111
0.18	4.47758004	0.17377286	46.32702984
0.22	5.30661798	0.26951624	36.75433522
0.26	5.88499162	0.37897578	29.18335798
0.30	6.22653324	0.49735297	23.19209125
0.38	6.32630562	0.74420074	14.68652622
0.46	5.89665058	0.98366934	9.34169835
0.54	5.19924000	1.20043003	5.97707156
0.62	4.42301194	1.38798071	3.85718832
0.70	3.68600178	1.54575902	2.52171250
0.86	2.54114132	1.78523800	1.15176602
1.02	1.90389106	1.95549826	0.61344815
1.18	1.66350460	2.09243949	0.40049408
1.34	1.67500492	2.22137860	0.31271098
1.50	1.81470498	2.35688878	0.27037070
1.82	2.17329958	2.66821433	0.21994441
2.14	2.40374068	3.02752279	0.17595280
2.46	2.44079822	3.40755925	0.13520379
2.78	2.32186608	3.78047749	0.10071257
3.10	2.11935604	4.12663307	0.07392931
3.74	1.58358290	4.70278438	0.03795192
4.38	1.11093634	5.12006110	0.01941230
5.02	0.75861404	5.40841008	0.01009136
5.66	0.51221250	5.60418540	0.00681363
6.30	0.34362100	5.73591491	0.00290225
7.58	0.15310642	5.88301683	0.00089329
8.86	0.06783632	5.94836369	0.00028969
10.14	0.03002402	5.97289938	0.00009788
11.42	0.01321300	5.99006357	0.00003396
12.70	0.00579816	5.99568309	0.00001205
15.26	0.00109610	5.99921414	0.00000158
17.82	0.00019803	5.99987404	0.00000021
20.38	0.00003547	5.99999110	0.00000003

5 DETERMINATION OF THE SELF-CONSISTENT FIELD FOR ATOMS

From the normalized atomic orbitals, there are several interesting quantities we can obtain. First, we can sum the quantities P^2 for all occupied orbitals, getting a radial charge density for the whole atom. We can sum the integrals of Table 5-1 to get the total charge inside a sphere of radius μx. Furthermore, since the radial charge density equals $4\pi r^2$ times the actual density, we can find this density $-\rho$, whose cube root gives the quantity proportional to the exchange potential V_X. We tabulate these three quantities in Table 5-2, using a mesh with twice the spacing of the presentation mesh. It is interesting to see that the charge density has a very high peak at the nucleus. It is also interesting to note, from the integrated function, that we must go out to about $x = 4$, or $r = 2$ bohrs, to get a large enough sphere to enclose five out of the six electrons of the atom.

Now let us turn to the main point of this chapter, the calculation of the self-consistent potential of Equations 1-2 and 2-4. The first term in the potential, $V_N(1) = 2Z/r_1$ from Equation 1-2, is obvious. The second term, $V_e(1) = \int \rho(r_2)(2/r_{12})\,dv_2$ of Equation 1-2, can be rewritten for a spherical distribution of charge in the form

$$V_e(1) = \frac{2}{r_1} \int_0^{r_1} 4\pi r_2^2 \rho(r_2)\,dr_2 + \int_{r_1}^{\infty} \frac{2}{r_2} 4\pi r_2^2 \rho(r_2)\,dr_2 \qquad (5\text{-}5)$$

This expresses the familiar fact that the charge within a sphere of radius r_1, equal to $\int_0^{r_1} 4\pi r_2^2 \rho(r_2)\,dr_2$, exerts a potential at radius r_1 as if it were all concentrated at the center of the sphere, whereas the charge in a spherical shell of radius from r_2 to $r_2 + dr_2$ produces a constant potential at interior points equal to the potential just outside the shell.

The integral in the first term of Equation 5-5 is already given in Table 5-2. The integral in the second term can be immediately computed from Simpson's rule, after constructing a table of functions $\sum P^2/x$. The third term in Equation 1-2 or 2-4 becomes in our case

$$V_X = 6(3/8\pi)^{1/3}[-\rho(1)]^{1/3} \qquad (5\text{-}6)$$

since we must set $\alpha = 1$ as Herman and Skillman have done, and we must set $\rho\uparrow$, the charge density with spin up, equal to $\rho/2$, where ρ is the total charge density. The charge density has already been given in Table 5-2. We

Table 5-3 Quantities $-V_e$, V_X, $-V_e - V_x$ computed from Herman and Skillman's wave functions, compared with $-V_e - V_x$ from Herman and Skillman's tabulated values U

x	$-V_e$	V_X	$-V_e - V_x$ from P's	From U's (Table 4-1)
0.02	29.89747683	14.47772090	15.41975593	15.418019
0.04	29.75961464	13.92429484	15.83531980	15.8367188
0.06	29.54849011	13.39189195	16.15659815	16.1569013
0.08	29.27707745	12.88157091	16.39550654	16.3970381
0.10	28.95655950	12.38968951	16.56686999	16.5682127
0.14	28.20549696	11.46439600	16.74110096	16.7427297
0.18	27.35719847	10.61019098	16.74700749	16.7480075
0.22	26.45702176	9.82233117	16.63469059	16.6349360
0.26	25.53769364	9.09543490	16.44225874	16.44298154
0.30	24.62243255	8.42477581	16.19765675	16.19877134
0.38	22.86314465	7.23465919	15.62848546	15.62866495
0.46	21.25178930	6.22186135	15.02992795	15.02983750
0.54	19.81272249	5.36135462	14.45136787	14.45154183
0.62	18.15450463	4.63306788	13.91197846	13.91204860
0.70	17.43511340	4.02108795	13.41402545	13.41459101
0.86	15.61940529	3.09671425	12.52269104	12.52148398
1.02	14.21949270	2.51017694	11.70931576	11.70887586
1.18	13.11599206	2.17759486	10.93839720	10.93796838
1.34	12.22123419	2.00520717	10.21602702	10.21569638
1.50	11.47436682	1.91028540	9.56408142	9.56360484
1.82	10.27248342	1.78326557	8.48921785	8.48862527
2.14	9.31629007	1.65542855	7.66086152	7.66032592
2.46	8.51577740	1.51626348	6.99951392	6.99887196
2.78	7.82661746	1.37448375	5.98423877	5.98637982
3.10	7.22413098	1.23989221	5.98423877	5.98637982
3.74	6.22392775	0.99278746	5.23114029	5.23157915
4.38	5.43603898	0.79396996	4.64206902	4.64242126
5.02	4.80675939	0.63840297	4.16835642	4.08853324
5.66	4.29740417	0.56006444	3.73733973	3.62622559
6.30	3.87950696	0.42139556	3.45811140	3.25784712

are, therefore, ready to compute the various terms in the potential, and we give them in Table 5-3.

The quantity $-V_e - V_X$ from P's in the table is made up from the calculations just mentioned. That is, it represents the part of the potential arising from the electrons, which we derive from the atomic orbitals at one stage in the iteration. If we have achieved self-consistency, as Herman and Skillman had done before tabulating their results, it should be equal to the same quantity calculated from the potential used in solving Schrödinger's equation. But this latter quantity has already been tabulated (for x up to 0.70) in Table 4-1. These values, extended to larger values of x, are given in the last column of Table 5-3. We see that we are getting agreement up to a few digits in the fifth significant figure throughout the table. In other words, we have done two things with this calculation: we have illustrated and verified the method of calculating the potential from the orbitals, and we have checked to see that Herman and Skillman had really achieved self-consistency.

The only place where there are discrepancies between the two calculations of $-V_e - V_X$ comes for x greater than 4.38. But this is just the place where Herman and Skillman introduced the Latter correction in their calculations. This means that for larger x's, their U's did not follow our prescription for self-consistent derivation of the potentials from the wave functions. Our values computed from the P's would be used in a further iteration, if we wished to eliminate the Latter correction used in Herman and Skillman's work. This correction would only affect the outermost electronic charge in the atom, and, in fact, its effect on the wave functions and energies is very small.

In addition to the quantities included in Table 5-3, it is interesting to consider the total potential acting on the electron, $V = V_N + V_e + V_X$, and the coulomb part of it, $V_N + V_e$. These are both tabulated in Table 5-4. The difference between the two columns is, of course, the quantity V_X given in Table 5-3. It is interesting to note that for small r, near the nucleus, the exchange term V_X is small compared to the total potential. On the other hand, for large x, at the outer periphery of the atom, the coulomb potential $V_N + V_e$ becomes very small, and the exchange energy decreases much more slowly with increasing x. A consequence of this is that in a molecule, in the bonding region between atoms the exchange term is rather more important than the coulomb term in determining the local potential acting on the electron.

Table 5-4 Values of total potential V acting on the electron, and its coulomb part $V_N + V_e$, for carbon

x	$V = V_N + V_e + V_X$	$V_N + V_e$
0.02	1216.053118	1201.575397
0.04	599.8988495	585.9745547
0.06	394.3334775	380.9415855
0.08	291.4707460	278.5891751
0.10	229.7260146	217.3363251
0.14	159.1817184	147.7173224
0.18	120.0821188	109.4719278
0.22	95.3169855	85.4946543
0.26	78.2855674	69.1901325
0.30	65.8993044	57.4745286
0.38	49.1856053	41.9509462
0.46	38.5123858	32.2905245
0.54	31.1585003	25.7971456
0.62	25.8128267	21.1797589
0.70	21.7702986	17.7492107
0.86	16.1173796	13.0206654
1.02	12.4376170	9.9274401
1.18	9.9344238	7.7568289
1.34	8.1644698	6.1592627
1.50	6.8560103	4.9457249
1.82	5.0440246	3.2607590
2.14	3.8487501	2.1933216
2.46	3.0130995	1.4968360
2.78	2.4080133	1.0335296
3.10	1.9585953	0.7187025
3.74	1.3538280	0.3610405
4.38	0.9807346	0.1867647
5.02	0.7379031	0.0995001
5.66	0.6141485	0.0540839
6.30	0.4513208	0.0299252
7.58	0.2940833	0.0095659
8.86	0.1986612	0.0031873
10.14	0.1390197	0.0028685
11.42	0.0960498	0.0003788
12.70	0.0678616	0.0001310
15.26	0.0344069	0.0000138

It is also interesting to note the fact that since the calculations of Herman and Skillman were made with $\alpha = 1$, whereas the present preferred value for carbon would be close to $\alpha = 0.75$, makes a very considerable change in the potential acting on the electron, particularly near the nucleus. The difference between V_X of Table 5-3, for small x, and $0.75V_X$, is of the order of magnitude of 3 rydbergs. In spite of this considerable difference in the potential between the calculations for $\alpha = 1$ and 0.75, the effect on the wave function is rather small. We have available calculations of the carbon atom made with $\alpha = 0.77$, tabulated in Note 1, and they show relatively small differences in the wave functions and total charge density, as compared to the Herman–Skillman calculations. The reason is that the exchange potential V_X is such a smooth function of r, as we see from Table 5-3, that it stays nearly constant over the region where an orbital is large. The main effect of the change of α is to change the orbital energies. Thus the maximum radial density of the 1s orbital comes at about $x = 0.34$. Here, as we see from Table 5-3, the value of V_X is of the order of 8 rydbergs. If we change the value of α by $\frac{1}{4}$, we might expect a change of energy of up to about 2 rydbergs. Actually, the calculation with $\alpha = 1$ has the 1s orbital coming about 1.2 rydbergs below that with $\alpha = 0.77$, the remaining part of the effect coming from the relatively small change of V_e from the one case to the other. Similarly, the maximum radial density of the 2s and 2p orbitals comes about at $x = 2.46$, where V_X is about 4.5 rydbergs. A change of this quantity by $\frac{1}{4}$ might be expected to change the energy by about 0.37 rydbergs. Actually, these orbitals for $\alpha = 1$ lie some 0.27 lower than with $\alpha = 0.77$. We shall consider the deeper significance of these energy changes later, but the interesting fact is that the shapes of the orbitals are not as sensitive to changes in α as we might have anticipated.

6 The total energy of an atom

So far our emphasis has been on the solution of Schrödinger's equation for the self-consistent field, Equation 2-4. Now we take up the other side of the problem, the determination of the total energy of an atom, from Equation 2-3. This involves the calculation of various integrals, and we find that the methods described in the preceding chapter allow us to compute all of these without trouble. Since these integrals are closely related to those we have met in the one-electron Schrödinger equation, let us first rewrite them and the related equations in a consistent notation. The one-electron Schrödinger equation for the atomic orbital u_i, which is given by Equation 3-1, is

$$\nabla^2 u_i = -(\varepsilon_i + V)u_i \qquad (6\text{-}1)$$

where

$$V = V_N + V_e + V_X$$
$$V_N = 2Z/r$$
$$V_e = \int \rho(r_2)(2/r_{12})\, dv_2$$
$$V_X = 6\alpha(3/8\pi)^{1/3}[-\rho(1)]^{1/3} \qquad (6\text{-}2)$$

In the preceding chapter we have seen how to compute V_e and V_X as functions of r; the computation of V_N is trivial. Hence if we rewrite the total energy of Equation 2-3 in terms of these integrals, we have the machinery for computing the energy. This total energy takes the form

$$\langle EX\alpha \rangle = -\sum(i)n_i \int u_i^* \nabla^2 u_i\, dv - \sum(i)n_i \int u_i^* V_N u_i\, dv$$
$$- \frac{1}{2}\sum(i)n_i \int u_i^* V_e u_i\, dv - \frac{3}{4}\alpha \sum(i)n_i \int u_i^* V_X\, dv \qquad (6\text{-}3)$$

equivalent to Equation 2-3. The total kinetic energy, the first term of Equation 6-3, can be found from the one-electron equations. If we multiply Equation 6-1 on the left by $-u_i^*$ and integrate, we have

$$-\int u_i^* \nabla^2 u_i\, dv = \varepsilon_i + \int u_i^* V_N u_i\, dv + \int u_i^* V_e u_i\, dv + \int u_i^* V_X u_i\, dv \qquad (6\text{-}4)$$

Thus we see that if the u_i's satisfy Equation 6-1, we can reduce the calculation of $\langle EX\alpha \rangle$ to the evaluation of the integrals $\int u_i^* V_N u_i\, dv$, $\int u_i^* V_e u_i\, dv$, and $\int u_i^* V_X u_i\, dv$, for the occupied orbitals. For carbon, where the only occupied orbitals are 1s, 2s, and 2p, this means only nine integrals.

Each of these integrands is the product of a charge density $u_i^* u_i$ and a function V_N, V_e, or V_X of r only. The integrations over angle, as in the problem of normalization, lead to the result that the integrals in question reduce to $\int_0^\infty P_{1s}^2 V_N\, dr$, $\int_0^\infty P_{1s}^2 V_e\, dr$, and so forth. We can at once multiply the functions

6 THE TOTAL ENERGY OF AN ATOM

P_{1s}^2 by V_N or V_e or V_X, as we have already been doing in the preceding chapter, and can then carry out the integrations over r by Simpson's rule. In other words, there is nothing in these integrals the reader cannot compute straightforwardly, using information we have already given. We therefore regard these quantities as being known, and tabulate them for the case of carbon, using the Herman–Skillman calculations we have been discussing. There is only one point that must be mentioned, on account of Herman and Skillman's use of the Latter correction. In the calculation of the kinetic energy integral, in Equation 6-4, the quantity $-V_e - V_X$ must be that in the last column of Table 5-3, rather than the next to last column. These two potentials differ slightly for the larger r values.

In Table 6-1 we give these integrals. We note first the effect of the Latter correction, which we mentioned above. The quantities on the right side of Equation 6-4, using the integrals from the table, prove to be 32.8028, 3.5306, and 2.6722 rydbergs, respectively for the 1s, 2s, and 2p orbitals, whereas the corresponding quantities, taking account of the Latter correction from the table are 32.8098, 3.5466, and 2.7344 rydbergs, respectively. This bears out our statement made earlier, that the effect of this correction is small, but must be considered.

Table 6-1 Atomic integrals for carbon, calculated from Herman and Skillman's orbitals. Integrals in rydbergs, and correspond to $\alpha = 1$

Kinetic energy integrals, $-\int u_i^* \nabla^2 u_i \, dv$	V_X integrals, $\int P^2 V_X \, dr$
1s, 32.8098 rydbergs	1s +6.2511
2s 3.5466	2s +1.4478
2p 2.7344	2p +1.2723

V_N integrals, $\int P^2 V_N \, dr$	One-electron energies $\varepsilon_i p$
1s +68.7078	1s −21.378
2s +11.4353	2s −1.2895
20 +9.6945	2p −0.6603

V_e integrals, $\int P^2 V_e \, dr$	
1s −20.7781	
2s −8.0630	
2p −7.6343	

Now let us use the integrals to compute the total energy of the atom. The total kinetic energy is 2(32.8098 + 3.5466 + 2.7344) rydbergs = 78.1815 rydbergs. The total potential energy is $-2(68.7078 + 11.4353 + 9.6945) + (20.7781 + 8.0630 + 7.6343) - 3/2\alpha(6.2511 + 1.4478 + 1.2723) = -143.1998 - 13.4568\alpha = -156.6566$ rydbergs if $\alpha = 1$. The total energy is then $78.1815 - 156.6566 = -78.4751$ rydbergs. As soon as we see this result, we look up in the tables to see how closely it agrees with the experimental value or with the Hartree–Fock value which is close to experiment. This latter quantity, from the calculations of J. B. Mann, is -75.31940 rydbergs. What is wrong?

To answer this question, we must go back to our derivation of the $X\alpha$ one-electron Schrödinger equations of Equation 2-4, from the $X\alpha$ total energy of Equation 2-3. This derivation was similar to that of the one-electron Hartree–Fock equations of Equation 1-2 from the Hartree–Fock total energy of Equation 1-6. In each case the variation method leads to the conclusion that the total energy of the ground state, if computed for orbitals that are not exact solutions of the one-electron equations, will lie somewhat above that found from the exact self-consistent solutions of the one-electron equations. This is the origin of the minimum property of the Hartree–Fock energy: any incorrect orbitals must lead to higher energy than the Hartree–Fock energy. How can it be that with our $X\alpha$ orbitals, we have got a lower energy than the Hartree–Fock value?

As we examine the question, we see that the apparent paradox is based on a misunderstanding. The variation method by which we derived the $X\alpha$ one-electron equations, Equation 2-4, from the $X\alpha$ total energy of Equation 2-3, was based on the assumption that the parameter α was held constant during the variation. All that it states is that with a fixed α, the orbitals that are exact solutions of Equation 2-4 will have a lower value of total energy of Equation 2-3 than would be found from any other orbitals, all computations using the same fixed α. On account of the exchange term of Equation 2-3, or Equation 6-3, proportional to α and with a negative coefficient, calculations with larger and larger values of α are bound to give lower and lower total energies, without limit. In other words, α cannot be chosen at will, or determined from a variational scheme. It must be fixed in some other way, once for all, and once it is given, then we can use the variational method to determine the best orbitals.

We believe that the values of α determined by Schwarz, as described in Chapter 2, give the best values to use. Thus specifically for carbon, for which Schwarz's α is 0.75847, the Hamiltonian $\langle EX\alpha \rangle$ for total energy given by Equation 2-3 or 6-3, using this value $\alpha = 0.75847$, has substantially the

6 THE TOTAL ENERGY OF AN ATOM

same properties as the Hartree–Fock Hamiltonian of Equation 1-6. Specifically, if we were to compute this energy of Equation 6-3, using $\alpha = 0.75847$ and using any arbitrary set of spin orbitals, the energy would lie above the Hartree–Fock value. The lowest energy would come for spin orbitals satisfying Equation 6-1 or 2-4, with $\alpha = 0.75847$. These spin orbitals would be very much like the Hartree–Fock spin orbitals, and the energy of Equation 6-3 would be very nearly as low as the Hartree–Fock values. Furthermore, for this lowest energy the kinetic and potential energy would obey the virial theorem, which for an isolated atom states that the total energy must be the negative of the kinetic energy or the potential energy must be twice the negative of the kinetic energy.

To test these statements, we could regard the Herman–Skillman orbitals computed for $\alpha = 1$ and using the Latter correction as an arbitrary set of orbitals, not satisfying Equation 6-3 with $\alpha = 0.75847$. The energy should lie above the Hartree–Fock value, and the kinetic and potential energies should not satisfy the virial theorem. Let us verify this expectation. The energy we wish to compute is identical with what we found earlier, leading to the value -78.4751 rydbergs, except that we earlier took the exchange term in the total energy to be -13.4568α, with $\alpha = 1$. If instead we had taken the exchange to be -13.4568α with $\alpha = 0.75847$, or -10.2066 rydbergs, we should have found a total energy of $78.1815 - 143.1998 - 10.2066 = -75.2249$ rydbergs. This, as we expected, is slightly above the Hartree–Fock value of -75.3194 rydbergs. The potential energy would have been $-143.1998 - 10.2066$ rydbergs $= -153.4064$ rydbergs, rather far from $-2(\text{kinetic energy}) = -2(78.4751) = -156.9502$ rydbergs, demanded by the virial theorem.

We have, of course, used the Herman–Skillman orbitals only as a pedagogical example, without any thought that they were very close to the correct orbitals. Schwarz and Connolly have found that for orbitals computed with $\alpha = 0.75847$, the energy of Equation 9-3 is -75.3087 rydbergs, showing that the energy we have found from the Herman–Skillman orbitals is above this best $X\alpha$ energy by only $-75.2249 + 75.3087 = 0.0838$ rydbergs. This indicates that very considerable deviations of orbitals from the correct $X\alpha$ ones produce only a quite small change in the total energy. The changes in Hartree–Fock energy, determined by Equation 1-6, when comparable changes are made in the orbitals, are of the same small order of magnitude.

It is interesting to see how small are the changes in the orbitals, in going from one of these methods of calculation to another. We have mentioned that in Note 1 we give the Herman–Skillman orbitals corresponding to

$\alpha = 1$. But we also give two other sets of orbitals: Mann's Hartree–Fock orbitals and a set of $X\alpha$ orbitals determined for $\alpha = 0.77$ (unfortunately we do not have a set calculated for exactly the value 0.75847). The reader can see that the three sets of orbitals are remarkably similar. Furthermore, he can verify his understanding of the various calculations which we have described by carrying them through for these other orbitals. He will find the same energy -75.3087 rydbergs for the total $X\alpha$ energy computed from the orbitals with $\alpha = 0.77$ which we have quoted above from Schwarz and Connolly. Their deviations from those for $\alpha = 0.75847$ are so small that the deviation of energy, a second-order quantity, is negligible.

Before leaving these points, it should be mentioned that, in fact, the calculations for the carbon atom are an example of a method used for an atom with a partially filled outer shell. The carbon atom has only two $2p$ electrons, whereas the complete $2p$ shell would have six electrons. The method used in place of the Hartree–Fock method, which is based on a single-determinant wave function, is called the hyper–Hartree–Fock method, and it gives an energy equal to the weighted mean of the various multiplet energies coming from the partially filled shell. This method has been described by the author, it is used by Mann, and one can show that the energy given by the $X\alpha$ method should agree with this weighted mean of the multiplet energies. It is this experimental weighted mean that agrees fairly well with the Hartree–Fock value -75.3194 rydbergs which we have quoted.

An additional point should also be mentioned in connection with Schwarz's calculations. He not only calculated the value 0.75847 for α for carbon, but also computed the quantity $\partial \langle EX\alpha \rangle / \partial \alpha$, the change in the quantity of Equation 6-3 with α, where now we compute $\langle EX\alpha \rangle$ using the variable α, rather than the fixed value 0.75847. He found this derivative to be -13.2 for carbon. Thus if one computed Schwarz's estimate of the energy to be found for carbon for $\alpha = 1$, one would have found $-75.3087 - 13.2(1 - 0.75847) = -78.4969$ rydbergs. Schwarz had no expectation of using his linear relation between energy and α over such a wide range of α as this, since his calculations were all for α near 0.75. But it is gratifying that our computed value of -78.4751 rydbergs agrees so well with Schwarz's estimate.

One would think at this point that everything was in order, and that if we determine α according to Schwarz's procedure, the $X\alpha$ method would have the same variational behavior as the Hartree–Fock method. However, quite a different story emerges as soon as we look at the one-electron energies

ε_i. It was largely to get good values of these energies, agreeing with optical excitation and ionization energies, that the original value $\alpha = 1$ was proposed. Herman and Skillman in their book give impressive graphs illustrating the good agreement they get between theory and experiment in this way, throughout the periodic table. They have found that if α deviates only slightly in either direction from unity, this agreement is greatly diminished. And yet we are proposing $\alpha = 0.75847$ for carbon. This paradox was very troublesome for several years, until the concept of the transition state was introduced, to interpret the optical excitation and ionization energies. In our study of the $X\alpha$ method, we must treat this aspect of the problem, and we take it up in the next chapter.

7 Excitations and the transition state

In the Hartree–Fock method, one has an important theorem called Koopmans' theorem, relating to the ionization energy of an atom or molecule. The theorem is the following: if ground-state orbitals u_i satisfy the Hartree–Fock equations, Equation 1-7, then the one-electron energy ε_i equals the energy of the atom, given by Equation 1-6, minus the energy of the ion formed when the occupation number of the ith orbital, which was $n_i = 1$ in the atom, is changed to $n_i = 0$. The theorem can be proved very simply, by subtracting the energies of atom and ion as given in Equation 1-6 and directly comparing with Equation 1-7. It is assumed that the orbitals u_i of the atom are still used in the calculation of the energy of the ion. It is Koopmans' theorem that has led scientists to identify the one-electron energies of the Hartree–Fock method with ionization energies. It is this identification that Herman and Skillman used in the comparison between their calculated energies and experiment, as we have mentioned at the end of the preceding chapter.

There is one point in which the ionization energies as given by Koopmans' theorem must be in error. If the u_i's satisfy the Hartree–Fock equations for the atom, they will not satisfy the equations for the ion. Hence the energy of the ion, as calculated from Equation 1-6 using the u_i's appropriate for the atom, must be too high, and Koopmans' theorem will give too large a value for the energy required to remove the electron from the atom. These errors are not very large, but they are appreciable.

The type of agreement we can expect is shown in Table 7-1. Here in the first column we see the experimental ionization energies of the $1s$, $2s$, and $2p$ orbitals in the carbon atom in its ground state configuration $1s^2 2s^2 2p^2$, estimated by the author from optical and X-ray experiments. The energies of atom and ion are the averages of multiplets in the configuration, as mentioned in the preceding chapter. The second column gives the Hartree–Fock eigenvalues, from the work of Mann, which would equal the ionization energies if Koopmans' theorem were exactly applicable and if the Hartree–Fock method were exactly correct. We see that the Hartree–Fock $1s$ energy is appreciably too large, a result of the error arising from the use of Koopmans' theorem, but the agreement of $2s$ and $2p$ energies is very good.

Table 7-1 One-electron energies of carbon $1s$, $2s$, and $2p$ orbitals. First column, experimental, from J. C. Slater, *Phys. Rev.* **98**, 1039 (1955). Second column, Hartree–Fock, from Mann. Third column, eigenvalues of $\alpha = 1$ calculation, Herman–Skillman. Fourth column, eigenvalues of $\alpha = 0.77$ calculation, Schwarz and Connolly. Energies in rydbergs.

	Exp	HF(Mann)	$\alpha = 1$(HS)	$\alpha = 0.77$
$1s$	−21.6	−22.67682	−21.378	−20.21537
$2s$	−1.43	−1.42412	−1.2895	−0.99831
$2p$	−0.79	−0.81380	−0.6603	−0.38518

For comparison, we also show in Table 7-1 the eigenvalues of the $1s$, $2s$, and $2p$ orbitals in the $X\alpha$ method, first as computed from $\alpha = 1$ by Herman and Skillman, then from the Schwarz–Connolly calculation with $\alpha = 0.77$. We see that the Herman–Skillman values are not bad, the $1s$ agreeing well with experiment and the $2s$ and $2p$ being somewhat low, but the lowering of α to 0.77, as in the last column, decreases the energies considerably. In other words, we are finding the discrepancy described in the preceding chapter, where we said that changing α appreciably from unity removes the fairly good agreement that Herman and Skillman had found between their eigenvalues and experiment.

7 EXCITATIONS AND THE TRANSITION STATE

The explanation of the discrepancy between the $\alpha = 0.77$ values and experiment is that when we look into it, we find that Koopmans' theorem does not apply with the $X\alpha$ method, and there is no reason to try to identify the eigenvalues with the ionization energies. The result on which Koopmans' theorem is based is the statement

$$\langle EHF \rangle (n_i = 1) - \langle EHF \rangle (n_i = 0) = \varepsilon_i \qquad (7\text{-}1)$$

where it is assumed as stated earlier that the u_i's and n's aside from n_i do not change when n_i goes from one to zero. But in place of Equation 7-1, the corresponding formula in the $X\alpha$ method is

$$\frac{\partial \langle EX\alpha \rangle}{\partial n_i} = \varepsilon_i \qquad (7\text{-}2)$$

where now the E_i's are determined as functions of the n_i's, from Equation 2-4. In other words, the one-electron energy, which is given by a finite difference in the Hartree–Fock method, is given by a derivative in the $X\alpha$ method.

Equation 7-2 can be easily proved from a comparison of Equations 2-3 and 2-4. We must regard the n_i's as being continuous variables, and hence we must admit the possibility of fractional occupation numbers, between zero and unity. Then we start with Equation 2-3 for $\langle EX\alpha \rangle$ and differentiate with respect to n_i. The terms parallel those found in varying u_i, in going from Equation 2-3 to 2-4 by the variation method, and the result proves to be precisely as in Equation 2-4, or in integral form in Equation 6-4. These variation methods are discussed more in detail in Note 3.

If the energy $\langle EX\alpha \rangle$ were a linear function of the n_i's, the type of finite difference found in Equation 7-1 would equal the derivative of Equation 7-2. However, this is not the case. It proves to be desirable to expand $\langle EX\alpha \rangle$ in power series about the value $n_i = 1/2$. Then we have

$$\langle EX\alpha \rangle (n_i) = \langle EX\alpha \rangle (n_i = 1/2) + \frac{\partial \langle EX\alpha \rangle}{\partial n_i} \bigg|_{1/2} (n_i - 1/2)$$

$$+ \frac{1}{2} \frac{\partial^2 \langle EX\alpha \rangle}{\partial n_i^2} \bigg|_{1/2} (n_i - 1/2)^2 + \frac{1}{6} \frac{\partial^3 \langle EX\alpha \rangle}{\partial n_i^3} \bigg|_{1/2} (n_i - 1/2)^3 \cdots \qquad (7\text{-}3)$$

from which

$$\langle EX\alpha \rangle (n_i = 1) - \langle EX\alpha \rangle (n_i = 0) = \frac{\partial \langle EX\alpha \rangle}{\partial n_i} \bigg|_{1/2} + \frac{1}{24} \frac{\partial^3 \langle EX\alpha \rangle}{\partial n_i^3} \bigg|_{1/2} \cdots$$

$$(7\text{-}4)$$

Thus if we compute the $X\alpha$ eigenvalue given in Equation 7-2 for a state in which the electron to be ionized is half missing, only half in the atom, and if we can disregard the third derivative term, we find that the $X\alpha$ eigenvalue gives a good estimate of the ionization energy.

We call this state, in which the electron to be ionized is half removed, a transition state. As we see, this method treats the two states, ground state and ionized state, in a symmetrical manner, rather than emphasizing the ground state at the expense of the ionized state, as is done in the Hartree–Fock method with use of Koopmans' theorem. When the transition-state $X\alpha$ method is used, the results, in general, prove to be more accurate than the Koopmans' theorem calculations. We have examples of the improvement made in this way in our later work.

It is not hard to see why there should be an improvement of the desired sort. Let us think of the carbon $1s$ orbital. In the transition state calculation we should be making a self-consistent calculation of a carbon ion containing only one and one-half $1s$ electrons, rather than two $1s$ electrons, as in the ground state. The nucleus will be less shielded by the electrons, and the one-electron energy will be lowered. This lowering is just about right to bring the $1s$ eigenvalue of the $X\alpha$ calculation with the value $\alpha = 0.75847$ into agreement with the experimental value (or with the $X\alpha$ eigenvalue for $\alpha = 1$). A similar explanation holds for the $2s$ and $2p$ orbitals, in which we should be making transition state calculations for atoms lacking one-half $2s$ or one-half $2p$ electrons.

The energies we get in this way are comparable to what would be found by making separate calculations of the $X\alpha$ energy for the atom and each of the ionized states. But there is a great advantage, particularly for much heavier atoms than carbon, in using the transition state technique. With that technique, the ionization energies are computed from the one-electron eigenvalues ε_i, whereas a calculation of the total energy $\langle EX\alpha \rangle$, and subtraction of the total energy of the atom and ion, involves the subtraction of very large quantities, making accuracy of calculation much more difficult.

Before we leave the question of the transition state, which we shall find particularly important for the molecular problem, we should mention that the case of ionization, which we have used as an example, is only a special case of a more general type of transition. Suppose we have an optical excitation, in which an electron is removed from one level that is occupied in the ground state, and is placed in a higher level that is empty in the ground state. Then we can set up a transition state, in which the occupation numbers in both the initial and final levels are set equal to 1/2. One can show equally simply that if we have found an $X\alpha$ self-consistent solution for this transition state, the excitation energy equals the difference of the $X\alpha$ eigenvalues between the initial and final electronic states.

There is a still further valuable feature of the transition state, suggested by Bohr's correspondence principle. If one finds the matrix component of dipole moment, leading to the optical transition probability between an initial and final state, and computes this using the transition state orbitals, one can get a very good value for the transition probability from this single matrix component. This is a great contrast to the usual method followed in Hartree–Fock calculations, in which everything is based on separate ground-state and excited-state calculations. In that case, one must carry out a summation over matrix components for each electron of the atom, each of which will have a different orbital in the ground state and excited state. This very elaborate calculation is not necessary if the transition state is used.

8 Spin polarization and the unrestricted self-consistent field

In almost all Hartree–Fock calculations that have been made for atoms, it is assumed that all spin orbitals of a given l value have the same radial dependence, irrespective of the value of the m associated with the orbit (generally called m_l) and the m_s associated with the spin, which is 1/2 for spin up, $-1/2$ for spin down. These are restrictions on the form of the function, and it is realized that one can approach the true Hartree–Fock solution more closely by removing the restrictions. It is convenience that has limited the use of the unrestricted Hartree–Fock method. If we are dealing with a spherical atom, we automatically come out with orbitals that depend on l but not on m_l, and if we are disregarding magnetic or multiplet effects, the spin orbitals will not depend on m_s.

Yet we know that the true Hartree–Fock solutions are not of this form. In the Hartree–Fock equations, Equation 1-7, each orbital has its own exchange term, and these terms do not have the form of a function of r times the orbital function. In Chapter 14 we come to methods of solving such a nonspherical problem. But we do not do too badly if we disregard these small differences in orbitals, as compared to those that would arise from a spherical potential independent of spin.

However, in the $X\alpha$ method, there is one aspect of these restrictions we can remove quite easily. This is the restriction which states that orbitals are independent of the spin. Calculations in which we allow different orbitals for different spins are generally called spin-polarized calculations, and they are of great use in problems of atomic multiplets and in discussing the magnetic behavior of molecules and solids. Let us illustrate this by the very simple case of the carbon atom.

The carbon atom in its ground state has the configuration $1s^2 2s^2 2p^2$. We have pointed out before that the $2p^2$ shell is incomplete, and it is known from multiplet theory that it leads to three multiplets, the ^3P, ^1D, and ^1S states. The experimental energy of the atom in these three states is -75.71369, -75.62106, and -75.51668 rydbergs, respectively. It is the weighted mean of these multiplets, giving them weights of 9, 5, and 1, respectively, namely -75.66968 rydbergs, which we must compare with Mann's Hartree–Fock calculation of this weighted mean, quoted in Chapter 6 as -75.31940 rydbergs, and with the $X\alpha$ value of -75.3087 rydbergs.

We might be interested in getting the best $X\alpha$ calculation possible of the true ground state, the ^3P with energy -75.71369 rydbergs. The characteristic of this state is that the spins of the two $2p$ electrons are parallel to each other. We can calculate it, in the Hartree–Fock method, from a single determinantal function in which both $2p$ orbitals, although they have different m_l's, both have spin up. Correspondingly, we could calculate it in the $X\alpha$ method by assuming both $2p$ orbitals to have spin up. In either case, we should get different radial equations for spin-up and spin-down orbitals, and they would have different one-electron energies.

Even after such a calculation were made, there would be no perfect agreement between the results of the self-consistent field and experiment. But spin orbitals computed taking account of spin polarization would give an energy for the ^3P multiplet that would agree better with experiment than if we had used the ordinary nonspin-polarized orbitals. Later, when we come to consider the formation of molecules, we shall often wish to use the spin-polarized calculations in the limit where the atoms were pulled far apart. For then the energy level would reduce to the correct multiplet level at large distances, whereas otherwise it would reduce only to the average energy of several multiplets. As we can see from our example of carbon, there can be a difference of several tenths of a rydberg between these energies.

It is harder, as we have mentioned, to remove the restriction that the orbitals are solutions of a central-field problem. This is particularly obvious in our molecular problems, which we start to discuss in the next chapter. In a diatomic molecule it is obvious that the potential around one atom will be quite different on the side facing the other atom from what it is on the side facing away from the other. And yet it is equally hard here to remove the approximation of spherical symmetry. As we shall see, for calculations of ordinary complexity we shall simply have to live with the consequences of assuming a spherical potential, or of using a restricted self-consistent-field method to determine the orbitals.

But we must always remember the consequence of the variation method. If we use the orbitals that have been determined from a restricted Hartree–Fock method, or restricted $X\alpha$ method, we can still build up many-electron wave functions and charge densities from these orbitals, and average the correct Hartree–Fock or $X\alpha$ Hamiltonian over these wave functions. If the errors in the wave function are regarded as small of the first order, the errors in total energy will be small of the second order. We have seen examples of sort of situation in Chapter 6, where we calculated the average $X\alpha$ energy, using the correct $\alpha = 0.75847$, for carbon orbitals determined with $\alpha = 1$. We found an error of only 0.08 rydbergs arising from the error in the orbitals. We base our treatment of molecules, which we take up next, on the hope that there will be a similarly small error in energy, if we use slightly incorrect orbitals for molecules and solids, but correctly average the Hamiltonian, either the Hartree–Fock or the $X\alpha$ Hamiltonian, over these somewhat incorrect wave functions.

9 The cellular and muffin-tin methods

In Chapter 1 we mentioned the work of Wigner and Seitz in introducing a cellular method for discussing the energy bands of such a crystal as sodium and their work in developing some of the ideas that have led to the $X\alpha$ method. It is time now to go more in detail into the cellular method. It occurred to Wigner and Seitz that it would be useful in their study of the sodium crystal, which has the body-centered cubic structure, to introduce cells in the form of polyhedra, one to an atom, with the nucleus at the center, and filling all space. These are cells in the form of cubes, with the eight corners cut off by planes midway between the atom and its eight nearest neighbors, so that the corner faces form regular hexagons. They made all their calculations inside a single cell, joining the solutions inside one cell onto those in the neighboring cells.

Wigner and Seitz then considered the nature of the potential (which for present purposes we may take to be the $X\alpha$ potential) acting on an electron inside one of these cells. They noticed that it would be very nearly spherically symmetrical, and they were able to estimate the very small deviations from spherical symmetry. It seemed reasonable to them, therefore, to treat the one-electron Schrödinger equation inside a cell like the spherically symmetrical problem we have been taking up in the preceding chapters. However, the wave function regular at the origin, inside a cell, must satisfy very different boundary conditions at the cell boundary from those that would lead to the eigenfunctions of an isolated atom.

In Chapter 3 we have considered solutions of the Schrödinger equation for a spherical potential. We may slightly generalize the work of that chapter, to a form adapted to the present purpose. For an arbitrary one-electron energy ε we can find solutions regular at the origin corresponding to each set of indices l, m of the spherical harmonics, which are to be identified with the quantum numbers l and m_l of the central-field problem. An arbitrary linear combination of this infinite set of solutions satisfies Schrödinger's equation and takes the form

$$u(\varepsilon,r,\theta,\phi) = \sum(lm) C_{lm} R_{lm}(\varepsilon,r) Y_{lm}(\theta,\phi) \qquad (9\text{-}1)$$

The methods of Chapters 3 and 4 show how the radial functions $R_{lm}(\varepsilon,r)$ can be computed. We have, then, a solution of Schrödinger's equation

$$\nabla^2 u = -(\varepsilon + V)u \qquad (9\text{-}2)$$

regular at the origin, containing an infinite number of arbitrary constants, the C_{lm}'s.

9 THE CELLULAR AND MUFFIN-TIN METHODS

If we were dealing with an isolated atom and used the methods of Chapters 3 and 4 to compute the radial functions out to infinity, we should find that no more than one of the infinite set of functions $R_{lm}(\varepsilon,r)$ would go to zero at infinite r, all others going infinite, and unless the energy ε was properly chosen, not even one such function would exist. The energies ε that lead to a single term regular at infinity would be the eigenvalues of the problem, and the functions would be the atomic eigenfunctions. But as we have stated, for a crystal the conditions are very different. We must use a different spherical expansion inside each cell, and we must choose the C_{lm}'s so as to make the functions from one cell join smoothly, with continuous function and continuous first derivative, everywhere over the bounding surface between one cell and the next.

A wave function in a periodic potential such as we have in a crystal must satisfy what is called the Bloch condition, namely, that it is multiplied by a factor $\exp(ik \cdot R_{pq})$ when one goes from a point in a cell around atom p to an equivalent point in a cell around atom q. Here k is the wave vector, R_{pq} the vector from the center of the pth cell to the center of the qth cell. We have a solution for each value of k, and the energy-band theory shows that the k's can be any vectors inside a so-called Brillouin zone in k space. We do not go into the details of the energy-band theory in this volume, but refer the reader to other texts. The problem is, given the wave vector k, find the energy for which we can satisfy the condition of continuity over the bounding surfaces of the cell. These conditions can be restated to refer to a single cell, since R_{pq} is also the vector from a point on one surface of a cell to a corresponding point on the opposite face. We then must demand that the function and slope at one of these two points must be multiplied by the Bloch factor $\exp(ik \cdot R_{pq})$ in going to the other equivalent point on the opposite face.

With the infinite set of arbitrary constants and the infinite number of boundary conditions to be satisfied in order to get continuity over each point in the surface of the cell, it seems reasonable that one can satisfy the conditions. However, to make the problem practical, one must obviously limit himself to a finite set of functions and arbitrary constants, and to satisfying the boundary conditions at a finite set of points on the surface of the cell. The first serious attempt to carry out this method was made by the author, for the case of sodium treated by Wigner and Seitz. In this work, eight points on the surface were used, namely, the midpoints between an atom and its eight nearest neighbors. These points are the centers of the eight hexagonal faces of the Wigner–Seitz cell. Eight functions were used, out of the infinite set of Equation 9-1. These functions consisted of one s function, for $l = 0$; three p functions, for $l = 1$; three d, for $l = 2$; and one f function, for $l = 3$. These were chosen to give functions of a suitable symmetry. Then there were eight conditions to be satisfied: four to get continuity of the function from one of the midpoints to the opposite midpoint and four to get continuity of the first derivative.

The equations between the C_{lm}'s to secure continuity are of the form that have no nonvanishing solutions unless the determinant of coefficients is zero. This gives a secular equation for determining the energy ε associated with the given k value. This secular equation can be solved, determining energy as a function of k, or the energy bands. Of course, since we have only a few C_{lm}'s, we get only a few energy bands, namely, the lowest, occupied bands and a few excited bands. Although these energy bands showed the proper general behavior, as a function of lattice spacing, subsequent work showed that they were not very accurate solutions of the problem. This work was with what is called the empty lattice test. If we set the potential V equal to zero in Equation 9-2, we can still use the method to determine energy bands, but we ought to find a single exponential, $\exp(ik \cdot r)$, as the solution, since it is an exact solution of the problem of zero potential, if k is properly chosen. The empty lattice test showed that the cellular method was giving a very poor approximation to the exponential and its energy as a function of k.

Consequently, further work with the method has been in the direction of satisfying boundary conditions at more and more points. Recent, very accurate calculations have used hundreds of points on the surface and hundreds of terms in Equation 9-1. Obviously, this is a cumbersome way to handle the situation. A useful suggestion was made in a paper of von der Lage and Bethe in 1947. They tried the effect of satisfying the boundary condition, not at the midpoint of each of the hexagonal faces but rather around a large circle in each of these faces. They determined the circles as the intersections of a sphere having the same volume as the cell, with the surface of the polyhedral cell. They found that if they satisfied the boundary conditions only around this circle, the resulting solutions, in fact, showed good continuity all over the surface of the cell. We come back to this von der Lage and Bethe method later.

The author in 1937, before this work of von der Lage and Bethe, had suggested another approach to the problem of satisfying the boundary conditions. He suggested in the first place that it would be simpler to satisfy

9 THE CELLULAR AND MUFFIN-TIN METHODS

boundary conditions over the surface of a sphere than over the complicated surface of a polyhedron. Therefore, he suggested using the spherical potential only inside a sphere inscribed in the polyhedron, using a constant potential in the region outside the spheres. One can make a simpler type of expansion in the regions of constant potential, fitting the solutions in those regions to the spheres with continuous function and derivative. This type of potential, spherical inside spheres, constant outside, came to be called a muffin-tin potential. Inside each sphere the expansion of the wave function was assumed to have the same form, given in Equation 9-1, as in the cellular method.

Outside the spheres, in the region of constant potential, it was assumed that the wave function was expanded as a linear combination of plane waves. It can be proved that for a given wave vector k, only plane waves of a form

$$\exp[i(k + K_i) \cdot r] \qquad (9-3)$$

where K_i is what is called a vector of the reciprocal lattice, will have the property of satisfying Bloch's condition. It was found that a good approximation to the correct wave function can be found using a quite limited number of K_i's. A secular equation for the energy was set up in the following way.

First, a single plane wave was joined to the spherical solution inside a sphere with continuous function, though discontinuous slope. This can be done easily, since the plane wave can be expanded in spherical coordinates around the center of the sphere (in terms of spherical harmonics of the angle) and spherical bessel functions of r. The properties of this expansion are described in Note 4. The combination of such a plane wave outside the spheres, supplemented by the spherical solutions inside the spheres, is called an augmented plane wave. Then one can find the matrix components of the energy between any two augmented plane waves, and in this way set up equations for the coefficients of the expansion in augmented plane waves that will exactly satisfy the Schrödinger equation for the muffin-tin potential, including continuity of the derivative of function over the surfaces of the sphere. It has proved to be quite possible to program this augmented plane wave method for digital computers, and as a result one can get very accurate solutions of the energy bands for the muffin-tin potential.

It is even possible, starting with the augmented plane waves, to solve the problem of energy bands using an exact, non-muffin-tin potential. One needs only to include in the matrix components of the Hamiltonian between the augmented plane waves terms arising from the deviation of the actual potential from the muffin-tin values. In a number of cases, estimates have been made of the effect of the non-muffin-tin corrections on the one-electron energies and on the total energy of the crystal. For a close-packed crystal like a metal, these corrections are small. Calculations of the total energy as a function of lattice spacing give compressibility, lattice spacing, and other properties in good agreement with experiment. It is very largely as a result of these calculations, which verify the reliability of the $X\alpha$ method which underlies the procedure, that it has seemed desirable to extend the general approach to the molecular problems which form the major topic of this book.

The augmented plane wave method (now abbreviated APW) is not directly applicable to the molecular problem, for we do not have the periodic potential that restricts one to a few of the plane waves of Equation 9-3. An expansion of the wave function of an electron in a molecule in plane waves converges so badly as to be useless. A more appropriate starting point for the calculations is an alternative treatment of the muffin-tin potential, suggested by Korringa, Kohn, and Rostoker (abbreviated KKR). This is the method underlying the multiple-scattering treatment (abbreviated MS-$X\alpha$) which has proved very successful for molecular problems.

In the KKR method, we work outside the muffin-tin spheres with solutions of the constant potential problem, expressed in spherical coordinates around one or another of the nuclei. A constant potential is, of course, a special case of the spherical potential we have been studying in the preceding chapters, so that we can apply the techniques we have learned to its solution. In the first place, the solutions must be of the form of functions of r times spherical harmonics of angle. The difference between this case and the general potentials we have been considering is that the radial differential equation has been studied for many years, and it leads to well-known functions for which we have analytic expressions. These radial functions are the spherical bessel or neumann functions, and the closely related spherical hankel functions. Properties of these functions are given in Chapter 10 and in Note 4, but we can describe their behavior here.

The behavior is completely different, depending on whether the energy ε of Equation 9-2 (where it is assumed that $V = 0$) is positive or negative. If ε is positive, we have for a particular ε value two independent solutions, one regular at the origin, the other not, and varying in a sinusoidal fashion

9 THE CELLULAR AND MUFFIN-TIN METHODS

at large r, related to each other like a sine to a cosine. These solutions are called the spherical bessel and neumann functions. They are used in the standard theory of scattering of waves. By combining the sine-like and cosine-like solutions, one can get incoming and outgoing progressive waves. The scattered wave from a center is made up out of the outgoing waves. These waves would come into the excited energy bands of a crystal, or as we have mentioned, they form scattered waves in a molecular problem.

If ε is negative, we again have two independent solutions, which can be written as spherical bessel and neumann functions of an imaginary argument. Again, one is regular at the origin, whereas the other one is not. But it is more useful to make linear combinations of these that behave like increasing or decreasing exponential functions of r at large r. In particular, there is a solution, called a spherical hankel function of the first kind, that is the decreasing function and that is adapted to represent the decreasing tail of an atomic orbital.

Korringa, Kohn, and Rostoker described the solution in the constant-potential region outside the spheres as a linear combination of functions of the type we have just described, centered on all the various nuclei in the crystal. The resulting sum is an exact solution of Schrödinger's equation for the constant potential. They then demanded that this sum join smoothly, with continuous function and slope, to the solutions inside each of the spheres. In this way, as in the APW method, they threw their boundary conditions into those over the surface of the spheres. This method, like the APW method, has been programmed for digital computers. In a classic calculation, Segall, working with the KKR method, and Burdick, working with the APW method, solved the energy bands of the copper crystal for an identical muffin-tin potential. To the limit of accuracy of their calculations, which was very high, their results for the energy bands agreed exactly. Hence we may consider that the two methods are essentially interchangeable when applied to crystals.

For the molecular problem, however, we have pointed out that the APW method is not appropriate, whereas the KKR method is readily adaptable. We merely form the solution in the constant-potential region as a sum of solutions of the type we have been describing, summed over the actual nuclei of the molecule. In this form, it has been programmed by Johnson and Smith for the digital computer. We describe this solution in the next chapter. It gives fairly reasonable results, with very rapid computations. But difficulties with the computed results have appeared, which indicate that the method as it stands is oversimplified.

The difficulty arises simply from the very large part of space in which a constant potential is used, rather than the actual potential, which is far from a constant. Similar difficulties have been observed in some crystals, in particular, for very loosely packed crystals like diamond, in which each atom has only four nearest neighbors, unlike the eight neighbors in the body-centered cubic structure or twelve in the face-centered cubic. Modifications of the APW method prove to be required in the diamond case, to get a better description of these large empty spaces in which actually a number of the electrons are located. We can see this difficulty in diamond easily from calculations we have already made. The carbon-carbon distance in diamond is about 1.54 Å = 2.90 bohrs, so that the radius of a nonoverlapping sphere in a muffin-tin potential would be about 1.45 bohrs, equivalent to an x value of $1.45/\mu = 2.98$. But in Table 5-2 we found that the number of electrons in a carbon atom in a sphere of radius corresponding to $x = 2.78$ was 3.78, or for radius 3.10 it was 4.13. In other words, for a sphere of radius 2.98, there would be only about four electrons inside the sphere, two outside.

The half carbon-carbon distance in the C_2 molecule is even smaller than in diamond, about 1.31 Å, corresponding to an x value of about 2.54. In this case, from Table 5-2, we should find only about 3.5 electrons inside each of the two atomic spheres, with 2.5 electrons outside. It is obvious that the muffin-tin method, with nonoverlapping spheres, is impracticable for diatomic molecules. With this in mind, Johnson and numerous other workers have suggested that the best way to retain this general approach is to make the spheres a good deal larger and to allow them to overlap. Remarkably enough, the computer programs do not have to be modified to make them adaptable to this overlapping-sphere case. The molecular orbitals obtained in this way prove to be very good approximations to the true orbitals, and the energy values also appear to be good. But many of the mathematical steps Johnson and his colleagues have taken appear to be unwarranted approximations. The point of view we take in the subsequent parts of this book is that the multiple-scattering, overlapping-sphere $X\alpha$ method has some very desirable properties, but that it would have to be examined much more critically than it has been in the existing literature before we could believe in its quantitative accuracy. We start this discussion in the next chapter.

10 The multiple-scattering method, weak interaction

The multiple-scattering method as originated by Johnson and Smith was dealing with nonoverlapping spheres, and it included not only these spheres and the interatomic region between them but also an outer sphere, surrounding the whole molecule, in which the wave function was again expanded in spherical harmonics and radial functions. In the discussion of this chapter, we do not use this last feature. Consequently, although our treatment leads to the same final formulas as those of Johnson and Smith, the derivation can be made simpler.

The problem we take up in this chapter is conceptually very simple. It is that in which the potential is spherically symmetrical within nonoverlapping atomic spheres, but constant and equal to zero in the space outside these spheres and out to infinity. There are very few actual problems to which this treatment is adapted: those in which the atoms are so far apart in proportion to their size that there is almost no electronic charge outside the spheres. For example, we could use this method to discuss a loosely bound molecule formed of inert gas atoms held together only by Van der Waals forces, so that they were very far apart. As we pointed out at the end of the preceding chapter, the method would be unsuited to such a molecule as C_2, where the atoms are much closer together, with much charge outside the spheres. But even though the method as we describe it is not suited to atoms that are tightly bound together, with strong interaction, nevertheless brings out important points in the mathematical discussion, which will guide us in further developments to be presented in later chapters.

We follow Johnson and Smith in referring to the regions inside the spheres as region I and outside space as region II. In region II, as we mentioned in the preceding chapter, Schrödinger's equation is the ordinary wave equation,

$$\nabla^2 u = -\varepsilon u, \qquad \frac{d^2 P}{dr^2} = -\left[\varepsilon - \frac{l(l+1)}{r^2}\right] P \qquad (10\text{-}1)$$

where $u = (P/r)Y_{lm}(\theta,\phi)$. We are dealing only with negative energies, or bound states, as found in the ground state of the molecule. The types of solution we wish to use for P/r, in spherical coordinates, are called $k_l^{(1)}(\kappa r)$

and $i_l(\kappa r)$, where $\kappa = \sqrt{-\varepsilon}$. These functions have the following properties:

$$\lim_{r \to 0} k_l^{(1)}(\kappa r) = (-1)^l \frac{1 \cdot 1 \cdot 3 \cdots (2l-1)}{(\kappa r)^{l+1}} \tag{10-2}$$

$$\lim_{r \to \infty} k_l^{(1)}(\kappa r) = (-1)^l \frac{\exp(-\kappa r)}{\kappa r} \tag{10-3}$$

$$k_{l+1}^{(1)}(\kappa r) = k_{l-1}^{(1)}(\kappa r) - \frac{2l+1}{r} k_l^{(1)}(\kappa r) \tag{10-4}$$

$$\frac{dk_l^{(1)}(\kappa r)}{d(\kappa r)} = \frac{1}{2l+1} \left[lk_{l-1}^{(1)}(\kappa r) + (l+1)k_{l+1}^{(1)}(\kappa r) \right] \tag{10-5}$$

$$-k_{-1}^{(1)}(\kappa r) = k_0^{(1)}(\kappa r) = \frac{\exp(-\kappa r)}{\kappa r} \tag{10-6}$$

$$k_1^{(1)}(\kappa r) = -\frac{\exp(-\kappa r)}{\kappa r}\left(1 + \frac{1}{\kappa r}\right) \tag{10-7}$$

$$k_2^{(1)}(\kappa r) = \frac{\exp(-\kappa r)}{\kappa r}\left[1 + \frac{3}{\kappa r} + \frac{3}{(\kappa r)^2}\right] \tag{10-8}$$

$$k_3^{(1)}(\kappa r) = -\frac{\exp(-\kappa r)}{\kappa r}\left[1 + \frac{6}{\kappa r} + \frac{15}{(\kappa r)^2} + \frac{15}{(\kappa r)^3}\right] \tag{10-9}$$

$$k_4^{(1)}(\kappa r) = \frac{\exp(-\kappa r)}{\kappa r}\left[1 + \frac{10}{\kappa r} + \frac{45}{(\kappa r)^2} + \frac{105}{(\kappa r)^3} + \frac{105}{(\kappa r)^4}\right] \tag{10-10}$$

$$\lim_{r \to 0} i_l(\kappa r) = \frac{(\kappa r)^l}{1 \cdot 3 \cdot 5 \cdots (2l+1)} \tag{10-11}$$

$$\lim_{r \to \infty} i_l(\kappa r) = \frac{\exp(\kappa r)}{\kappa r} \tag{10-12}$$

$$i_l(\kappa r) = \frac{(\kappa r)^l}{1 \cdot 3 \cdot 5 \cdots (2l+1)} \left[1 + \frac{(\kappa r)^2}{2(2l+3)} + \frac{(\kappa r)^4}{2 \cdot 4 \cdot (2l+3)(2l+5)} + \cdots \right] \tag{10-13}$$

$$\frac{di_l(\kappa r)}{d(\kappa r)} = \frac{1}{2l+1} \left[li_{l-1}(\kappa r) + (l+1)i_{l+1}(\kappa r) \right] \tag{10-14}$$

$$i_l(\kappa r) \frac{dk_l^{(1)}(\kappa r)}{d(\kappa r)} - k_l^{(1)}(\kappa r) \frac{di_l(\kappa r)}{d(\kappa r)} = \frac{(-1)^{l+1}}{\kappa r^2} \tag{10-15}$$

10 THE MULTIPLE-SCATTERING METHOD, WEAK INTERACTION

We have given the formulas for the functions in the form appropriate to the uses we shall make of them. That is, these formulas for $k_l^{(1)}(\kappa r)$ include the decreasing exponential, $\exp(-\kappa r)$, so that they all go to zero at infinity. The formulas for $i_l(\kappa r)$, being regular at the origin, can be expanded in power series in r. The recursion formula of Equation 10-4 allows us to generate as many of the functions $k_l^{(1)}$ as we desire.

In the region II, which extends out to infinity, we can build up the solution of Schrödinger's equation by superposing functions of the form $k_l^{(1)}(\kappa r)Y_{lm}(\theta,\phi)$ around each nucleus. This function will be bound to go to zero at infinity, and being a linear combination of solutions of Equation 10-1, all for the same energy, it will be a general solution of that equation, going to zero at infinity. That is, as the general solution in region II, we may write

$$\Psi_{II} = \sum (b,l,m) A_{lm}^b k_l^{(1)}(\kappa|r_b|) Y_{lm}(\hat{r}_b) \qquad (10\text{-}16)$$

where b denotes the bth atom $|r_b|$ is the distance from the bth nucleus to the point where the wave function is being computed, and in the spherical harmonic $Y_{lm}(\hat{r}_b)$ means the spherical harmonic for the angle of the vector r_b, using the bth nucleus as origin. We must now apply the boundary conditions to make this function continuous, with continuous derivative, to the solution of the type of Equation 9-1 inside the sphere over all the spherical surfaces bounding the atomic spheres.

The type of expansion of the wave function given in Equation 10-16 is one possible way of describing it. But there is an alternative form of expansion which is more convenient for treating the continuity of the wave function over the surface of the spheres. Suppose we wish to expand within a cell containing the particular atom a. There will then be two types of terms in Equation 10-16: those for which $b = a$, the functions representing the tails of the atom a, and those for which $b \neq a$, representing all the other atoms. Let us try to expand these tails of atoms b, within the cell a, in terms of functions of r_a and spherical harmonics of the angle r_a. It would, of course, be possible to expand these tails of other atoms in power series in x, y, z about the nucleus a. This could be done by Taylor's series, but there is a more straightforward way to do it. This is a result of a general theorem by which we can expand a function of the form $k_l^{(1)}(\kappa|r_b|)Y_{lm}(\hat{r}_b)$ as a linear combination of functions $i_{l'}(\kappa|r_a|)Y_{l'm'}(\hat{r}_a)$ around the center a. Since the functions $i_{l'}(\kappa|r_a|)$ can be expanded in power series by Equation 10-13, and since we show in Note 4 that the spherical harmonics $Y_{l'm'}(\hat{r}_a)$ can also be expanded in power series in the x, y, z coordinates around nucleus a, the functions $i_{l'}(\kappa|r_a|)Y_{lm}(\hat{r}_a)$ can be expanded in power series in x, y, z, or vice versa.

The theorem we have mentioned above is

$$k_l^{(1)}(\kappa|r_b|)Y_{lm}(\hat{r}_b)$$
$$= \sum_{l'm'L} 4\pi(-1)^{l'+1} I_L(l'm';lm) k_L^{(1)}(\kappa|r_{ab}|) Y_{L,m'-m}^*(\hat{r}_{ab}) i_{l'}(\kappa|r_a|) Y_{l'm'}(\hat{r}_a) \quad (10\text{-}17)$$

Here

$$r_{ab} = r_a - r_b \quad (10\text{-}18)$$

and

$$I_L(lm;l'm') = \int Y_{L,m-m'}(r) Y_{lm}^*(r) Y_{l'm'}(r)\, d\Omega \quad (10\text{-}19)$$

where the integration over $d\Omega$ is over all solid angles. The integrals over a product of three spherical harmonics in Equation 10-19 are the same ones met in the definition of the well-known coefficients $c^k(lm;l'm')$ in the theory of atomic spectra. We discuss these coefficients further in Chapter 14 and Note 6. In Equation 10-17 we have the contribution of the bth atom to the summation of Equation 10-16 in terms of an expansion around the ath nucleus.

Let us now write the whole function Ψ_{II} in this form. We must then sum over all nuclei b except the ath and include separately the term for nucleus a. We have

$$\Psi_{II} = \sum_{lm} \Bigg[A_{lm}^a k_l^{(1)}(\kappa|r_a|) Y_{lm}(\hat{r}_a)$$
$$+ \sum_{b \neq a, l'm'L} A_{lm}^b 4\pi(-1)^{l'+l} I_L(l'm';lm) k_L^{(1)}(\kappa|r_{ab}|) Y_{L,m'-m}^*(\hat{r}_{ab}) i_{l'}(\kappa|r_a|) Y_{l'm'}(\hat{r}_a) \Bigg]$$
$$(10\text{-}20)$$

In the second term we interchange the names of the symbols l,m with l',m'. Then we have

$$\Psi_{II} = \sum_{lm} Y_{lm}(\hat{r}_a) \Bigg[A_{lm}^a k_l^{(1)}(\kappa|r_a|)$$
$$+ \sum_{b \neq a, l'm'L} A_{l'm'}^b 4\pi(-1)^{l+l'} I_L(lm;l'm') k_L^{(1)}(\kappa|r_{ab}|) Y_{L,m-m'}^*(\hat{r}_{ab}) i_l(\kappa|r_a|)$$
$$= \sum_{lm} Y_{lm}(\hat{r}_a) \Bigg[A_{lm}^a k_l^{(1)}(\kappa|r_a|) - \sum_{b \neq a, l'm'} (-1)^{l'} \kappa^{-1} G_{lm;l'm'}^{ab}(\hat{r}_{ab};\varepsilon) A_{l'm'}^b i_l(\kappa|r_a|)$$
$$(10\text{-}21)$$

where

$$G^{ab}_{lm;l'm'}(r_{ab};\varepsilon) = (1 - \delta_{ab})4\pi(-1)^{l+1} \sum_L I_L(lm;l'm') k_L^{(1)}(\kappa|r_{ab}|) Y^*_{L,m-m'}(\hat{r}_{ab})$$
(10-22)

We can then rewrite Equation 10-21 in the form

$$\Psi_{II} = \sum_{lm} Y_{lm}(r_a)\Psi_{II}(lm)$$
(10-23)

where $\Psi_{II}(lm)$ is the bracket multiplying $Y_{lm}(\hat{r}_a)$ in Equation 10-21. This gives a solution of Equation 10-1, holding inside the sphere a and converging for a short distance outside.

We must now produce continuity of the wave function and its first derivative between the point immediately inside the sphere surrounding the ath nucleus, where the wave function is a linear combination of functions $R_l(\kappa|r_a|)Y_{lm}(\hat{r}_a)$, and a point immediately outside the sphere, where the function is given by Equation 10-21, convergent if r_a is only slightly greater than the sphere's radius. For continuity of function and derivative at the surface of the sphere, we make the logarithmic derivative of the term for each l, m continuous, as well as the function Ψ itself and its normal derivative. That is, at the radius of the ath sphere,

$$\frac{1}{\Psi_{II}(lm)} \frac{d\Psi_{II}(lm)}{d|r_a|} = \frac{1}{R_l^a(\kappa|r_a|)} \frac{dR_l^a(\kappa|r_a|)}{d|r_a|}$$
(10-24)

If we use a bracket notation for an expression called the Wronskian in mathematics, defined as

$$[A(x),B(x)] = A(x)\frac{dB(x)}{dx} - B(x)\frac{dA(x)}{dx}$$
(10-25)

we can rewrite Equation 10-24 in the form

$$[R_l^a(\kappa|r_a|), \Psi_{II}(l,m)] = 0$$
(10-26)

When we substitute Equation 10-21 into this expression, we have

$$A^a_{lm}[R_l^a(\kappa|r_a|), k_l^{(1)}(\kappa|r_a|)]$$
$$- \sum_{b \neq a, l'm'} \kappa^{-1} G^a_{lm,l'm'}(r_{ab};\varepsilon)(-1)^{l'} A^b_{l'm'} [R_l^a(\kappa|r_a|), i_l(\kappa r_a)] = 0 \quad (10\text{-}27)$$

We now divide both terms of this equation by $[R_l^a(\kappa|r_a|).i_l(\kappa|r_a|)]$ and find

$$A^a_{lm} \frac{[R_l^a(\kappa|r_a|), k_l^{(1)}(\kappa|r_a|)]}{[R_l^a(\kappa|r_a|), i_l(\kappa|r_a|)]} - \sum_{b \neq a, l'm'} \kappa^{-1} G^{ab}_{lm,l'm'}(r_{ab};\varepsilon)(-1)^{l'} A^b_{l'm'} = 0 \quad (10\text{-}28)$$

where we have such an equation for a equal to each one of the atoms and for each l, m for which the coefficients A_{lm}^a are of appreciable size.

In Equation 10-28 we have found the method of getting the coefficients A_{lm} of the wavelets in the region II. We need in addition the coefficients C_{lm} of Equation 9-1, for the expansion inside the atomic spheres. Let us derive the relation between the A's and the C's. We start with Equations 9-1 and 10-21. We use Equation 10-27 to express the summation over b, $l'm'$, which appears in Equation 10-21 in terms of A_{lm}^a. We demand that the summations of Equations 9-1 and 10-21, representing the solutions inside and outside the ath sphere, should be equal at the radius r_a of the sphere. This demands that the coefficients of the spherical harmonics $Y_{lm}(\theta,\phi)$ in both summations should be equal for each l and m as well as for each sphere. This leads to the equation

$$C_{lm}^a R_l^a(\varepsilon,|r_a|) = A_{lm}^a \{k_l^{(1)}(\kappa|r_a|) - \frac{[R_l^a(\varepsilon,|r_a|), k_l^{(1)}(\kappa|r_a|)]}{[R_l^a(\varepsilon,|r_a|), i_l(\kappa|r_a|)]} i_l(\kappa|r_a|) \quad (10\text{-}29)$$

By use of Equation 10-15 we can then show that

$$A_{lm}^a = C_{lm}^a \frac{r_a^2}{(-1)^l} [R_l^a(\varepsilon,|r_a|), i_l(\kappa|r_a|)] \quad (10\text{-}30)$$

which is the desired relation between the A's and the C's.

In the foregoing discussion, in particular in Equation 10-28, we have the fundamental equations of the multiple-scattering cluster method. As always with the KKR or scattered-wave method, the coefficients A_{lm}^b determined by these equations are the coefficients of the expansion of the wave function in the interatomic region II, by Equation 10-16, in terms of the scattered wavelets emitted from each atom. As we have seen, for negative energy these wavelets, instead of being sinusoidal as they would be for positive energy, are falling off exponentially as we go away from the bth nucleus. We have as many simultaneous linear equations in Equation 10-28 as we have wavelets. In many actual cases this means that only small values of l need be used, for this expansion in scattered wavelets is rapidly convergent. In general, the equations will have no nonvanishing solutions, and it is only if the determinant of coefficients of Equation 10-28 is zero that we get solutions. Since the energy ε is entering into the coefficients, we have solutions only for certain eigenvalues of the problem, which is the method by which the energy levels are found. Once the determinant is zero, Equation 10-28 can be solved for the coefficients A_{lm}^a, and hence for the wave function in region II. In turn, from Equation 10-30 we can find the wave function inside each sphere.

Though the problem appears at first sight quite complicated, it has been very efficiently programmed for the digital computer, and in practice the eigenvalues and eigenfunctions can be determined easily. One of the advantages of the method of procedure is that the quantities of Equation 10-22 depend only on the geometry and can be computed once and for all for a given atomic arrangement. It is only in the quantity in the first term of Equation 10-28 that the particular properties of the atoms enter, through the logarithmic derivatives of the functions R_l. We have set up the problem as if there were no regularity in the arrangement of the atoms, and one of the strengths of the method is that it can be used for molecules of a complicated form. However, if we have a regular crystal and are looking for a Bloch function as the solution, we can write the coefficients A_{lm} for any atom in the crystal as the coefficients for the corresponding atom in the central unit cell times a Bloch factor $\exp(ik \cdot R)$, so that actually we need only as many coefficients as would be required for the contents of this unit cell. When we put in the condition of periodicity, we are led to the equations of the KKR or scattered-wave method. On the other hand, equally simple solutions hold when we have one or a few perturbed atoms, embedded in an otherwise periodic crystal.

A remarkable feature of the mathematical analysis we have given in this chapter is that it carries over without change to treatments that have been quite successful in handling problems of strong interactions between atoms, as well as the weak interactions to which the treatment is directly adapted. However, in this case, we meet the many uncertain aspects of the problem which we mentioned at the end of Chapter 9. Consequently, in the next chapter we pause and examine some of these difficulties, and look for ways in which they can be overcome.

11 Qualitative nature of molecular orbitals

Before we go on to look at the shortcomings of the multiple-scattering method as in use at present, we should become familiar with the commonly held views regarding molecular orbitals. We have mentioned that these views are mainly based on the LCAO, or linear combination of atomic orbitals, approach. As we have mentioned earlier, these methods have been used with great success for simple molecules; the difficulties become great only when we are dealing with very large molecules. Even the diatomic molecule can give us a great deal of information regarding the type of molecular orbital which we must compute, and consequently, the problem the extensions of the multiple-scattering method must expect to meet.

The author wrote in 1963 a text entitled *Quantum Theory of Molecules and Solids, Volume 1: Electronic Structure of Molecules*, (McGraw-Hill Book Co., New York). In this text an effort was made to present the best picture then available of the molecular orbital theory. Let us give approximately the view then current. As we have mentioned previously, molecular orbitals were generally built up as linear combinations of atomic orbitals. There was at the time no convenient tabulation of atomic orbitals (the book of Herman and Skillman was only published in 1963), and it was the style to build up the atomic orbitals as linear combinations of functions of the type $r^n \exp(-ar) Y_{lm}(\theta, \phi)$, as we have mentioned earlier; or somewhat later, of gaussian functions $r^n \exp(-ar^2) Y_{lm}(\theta, \phi)$. Many such functions had to be superposed to get an acceptable approximation to an atomic orbital. It was a major effort, from the information given in the published papers, to get any graphical or tabular information about the actual values of the molecular orbitals. However, it was the custom to give computed values of the one-electron energies of the various orbitals.

As an illustration, we shall discuss simple diatomic molecules as treated in Chapters 6 and 7 of the text quoted above. In particular, when we wish a single illustration, we choose C_2, partly because it has been the subject of much calculation, using the multiple-scattering method, by J. B. Danese and J. W. D. Connolly. It is with this molecule in mind that we have used the carbon atom as an example of methods of making atomic calculations.

In Table 6-1 of the quoted text, reproduced in Table 11-1, we gave the Hartree–Fock one-electron energies for homonuclear diatomic molecules from Li_2 to F_2, using the most accurate values then available. In Fig. 6-1, reproduced here in Fig. 11-1, these one-electron energies were plotted as a function of atomic number. They show that the C_2 molecule has five occupied molecular orbitals in its ground state, the structure being

$$(1\sigma_g)^2 (1\sigma_u)^2 (2\sigma_g)^2 (2\sigma_u)^2 (1\pi_u)^4.$$

The one-electron energies of these orbitals were respectively -22.6775, -22.6739, -2.0567, -0.9662, and -0.8407 rydbergs, as computed by

Table 11-1 One-electron energies for homonuclear diatomic molecules, computed by molecular-orbital self-consistent-field method. Energies in rydbergs. Occupied orbitals to left of lines, orbitals enclosed by lines half occupied. References given below.

Mol.	Ref.	$1\sigma_g$	$1\sigma_u$	$2\sigma_g$	$2\sigma_u$	$1\pi_u$	$3\sigma_g$	$1\pi_g$	$3\sigma_u$
Li$_2$	1	−4.8806	−4.8802	−0.3604	0.0580	0.1282	0.1834	0.3206	0.7230
	2	−4.8710	−4.8705	−0.3627	0.0551		0.2158		0.7918
Be$_2$	2	−9.4187	−9.4184	−0.8512	−0.4416		0.1110		1.0882
B$_2$	3	−15.3530	−15.3514	−1.3552	−0.6990	−0.6824	0.0172		1.2898
C$_2$	2	−22.6775	−22.6739	−2.0567	−0.9662	−0.8407	−0.0444	0.5295	2.2598
N$_2$	4	−31.4438	−31.4396	−2.9054	−1.4612	−1.1595	−1.0892	0.5459	2.2054
	2	−31.2941	−31.2885	−2.8421	−1.4274	−1.0908	−1.1110	0.6004	2.2454
O$_2$	5	−41.1902	−41.1854	−3.0438	−1.9564	−1.0998	−1.1128	−0.7888	1.4784
F$_2$	2	−52.7191	−52.7187	−3.2517	−2.7225	−1.2159	−1.0922	−0.9489	0.6848

From *Quantum Theory of Molecules and Solids*, Vol. 1 by John C. Slater. Copyright © 1963 by the McGraw-Hill Book Company, Inc. Used with permission of McGraw-Hill Book Company.

References

[1] E. Ishiguro, K. Kayama, M. Kotani, and Y. Mizuno, *J. Phys. Soc. Japan*, **12**:1355 (1957).
[2] B. J. Ransil, *Revs. Modern Phys.*, **32**:239, 245 (1960).
[3] A. A. Padgett and V. Griffing, *J. Chem. Phys.*, **30**:1286 (1959).
[4] C. W. Scherr, *J. Chem. Phys.*, **23**:569 (1955).
[5] M. Kotani, Y. Mizuno, K. Kayama, and E. Ishiguro, *J. Phys. Soc. Japan*, **12**:707 (1957).

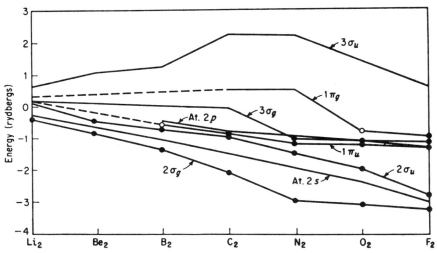

Figure 11-1 One-electron energies of homonuclear diatomic molecules. For comparison, 2s and 2p energies (experimental) of the separate atom^J. Occupied orbitals indicated by black circles, half-occupied by open circles, unoccupied without circles. From QUANTUM THEORY OF MOLECULES AND SOLIDS, Vol. 1 by Slater. Copyright © 1963 by McGraw-Hill, Inc. Used with permission of McGraw-Hill Book Company.

B. J. Ransil. These numbers compare with the Hartree–Fock atomic values of Mann, quoted in Table 7-1, namely -22.67682 Rydbergs for $1s$, -1.42412 rydbergs for $2s$, and -0.81380 rydbergs for $2p$. We note that the $1s$ atomic orbital and the molecular $1\sigma_g$, $1\sigma_u$ have almost precisely the same energy, whereas the molecular $2\sigma_g$ is definitely below the atomic $2s$, and the molecular $2\sigma_u$ and $1\pi_u$ slightly below the atomic $2p$.

Let us now remind ourselves how the molecular orbitals are formed from the atomic orbitals, according to the conventional views of the chemists. The molecular $1\sigma_g$ and $1\sigma_u$ are essentially normalized sums and differences of atomic $1s$ orbitals on the two atoms. The $2s$ orbitals of the two atoms and the $2p$ orbitals with $m_l = 0$ (wave functions independent of rotation around the internuclear axis) will lead to two σ_g orbitals, made up out of sums of atomic orbitals on the two centers, and two σ_u orbitals, made up out of the differences. However, as a result of hybridization, the lowest σ_g molecular orbital will be made up partly of $2s$, partly of $2p$ atomic orbitals, and similarly with the other molecular orbitals. The atomic $2p$ orbitals with $m = \pm 1$ will combine into a doubly degenerate $1\pi_u$, the linear combination of the atomic orbitals that is even on reflection in the midplane between the atoms (and consequently odd on inversion in the midpoint between the nuclei), whereas the $1\pi_g$ is the linear combination of atomic orbitals that is odd on reflection, even on inversion. Thus the linear combinations of these atomic orbitals lead not only to the occupied molecular orbitals, but also to $3\sigma_g$. $3\sigma_u$, and $1\pi_g$ orbitals, which are empty in the molecule, accounting for the four electrons missing from two neon shells, with their sixteen electrons. The $3\sigma_g$ is an excited level, lying not very far above the occupied orbitals, whereas the $3\sigma_u$ and $1\pi_g$ orbitals lie much higher in energy.

The interesting feature of this molecular orbital formation is the question of hybridization. Let us investigate this by looking into the case of complete hybridization, where we shall first make a combination of the $2s$ and $2p\sigma$ atomic orbitals, taken arbitrarily with equal coefficients, and then make symmetric and antisymmetric combinations of these on the two atoms. In Fig. 11-2 we make combinations of the radial functions R_s and R_p (taken equal to P_{2s}/x and P_{2p}/x for the atomic Herman–Skillman orbitals given in Note 1), such as we would have for $R_s + R_p \cos\theta$ for $\cos\theta = 1, 0, -1$, $\cos\theta$ being the angular function for $l = 1$, $m_l = 0$. The outer peaks of the two wave functions are approximately equal in magnitude (although opposite in sign with the convention used), so that $R_s + R_p$ is almost zero once we get outside the region near the nucleus, whereas $R_s - R_p$ shows a high (negative) peak.

11 QUALITATIVE NATURE OF MOLECULAR ORBITALS

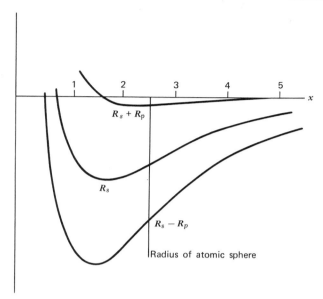

Figure 11-2 Linear combinations of radial wave functions R_s and R_p for Herman–Skillman 2s and 2p orbitals. Ordinate is arbitrary. Abscissa is distance from nucleus, $x = r/u$.

We may then expect that a linear combination of the functions $R_s - R_p$ on the two atoms, combined so that the peaks add, will have a large concentration of charge between the atoms and will have the lowest one-electron energy for its molecular orbitals. This is the interpretation we give for the low energy of the $2\sigma_g$ orbital. The $2\sigma_u$ is interpreted as coming from an antisymmetric combination of the hybrids like $R_s + R_p$, with opposite signs on the two nuclei.

In Fig. 11-3, we show these linear combinations of hybridized orbitals, plotted along the internuclear axis, leading to the $2\sigma_g$ and $2\sigma_u$ types of orbitals. They may be compared with the molecular orbitals for O_2, given in Fig. 6-4 of the text quoted above, and reproduced in Fig. 11-4. The agreement between our orbitals of Fig. 11-3, obtained with no calculation whatever, and those of Fig. 11-4, obtained by elaborate calculation, is striking. The O_2 calculations include the $3\sigma_g$ and $3\sigma_u$ orbitals as well, $3\sigma_g$ being occupied in O_2. though not in C_2, whereas $3\sigma_u$ is unoccupied both in C_2 and O_2. The two nuclei in Fig. 11-3 are assumed to come at $x = \pm 2.5$, which is about the correct distance for the molecule of C_2.

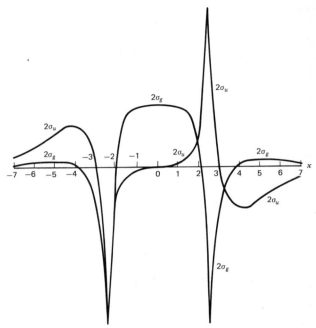

Figure 11-3 Approximate $2\sigma_g$ and $2\sigma_u$ molecular orbitals of C_2, made up from symmetric combination of $R_s - R_p$ of Fig. 11-2 and antisymmetric combination of $R_s + R_p$ as function of $x = R/u$. Nuclei are located at $x = +2.5$. Quantities plotted are $-2\sigma_g$ and $2\sigma_u$ where $2\sigma_u$ is $R_s + R_p$ on right-hand atom minus $R_s + R_p$ on left-hand atom.

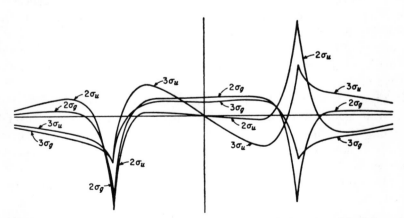

Figure 11-4 Molecular orbitals for O_2, along internuclear axis, for some of the lower states. From QUANTUM THEORY OF MOLECULES AND SOLIDS, Vol. 1 by Slater. Copyright © 1963 by McGraw-Hill, Inc. Used with permission of McGraw-Hill Book Company.

The large concentration of charge between the nuclei in the $2\sigma_g$ orbital and the lack of any charge in this region in the $2\sigma_u$ are clearly shown in the figures, as, of course, are the origin of their bonding and antibonding properties, respectively. We have not shown similar curves for the $1\pi_u$ orbital, since it is zero along the axis, and one must use more elaborate methods for describing it. However, wave function contours and charge densities obtained from the orbitals we have set up for all three types of orbitals show striking resemblance to those found by various elaborate computer methods. The most significant feature, as we next point out, is the large amount of charge outside the nonoverlapping atomic spheres.

The half distance between carbon atoms in the molecule, as we have mentioned earlier, is about 2.5 in x units; more accurately about 2.54. In Fig. 11-2 we have indicated this radius, and it is obvious how large are the tails of the wave functions extending outside the radius. This is really enhanced, since the charge between r and $r + dr$ is $4\pi r^2 R^2 dr$, or $4\pi P^2 dr$. The factor r^2 increases the contribution from larger values of r. In fact, from Table 5-2, we see that the total charge in the carbon atom, outside a sphere of radius 2.46 in x units, is $6 - 3.407 = 2.593$ electrons, whereas outside a sphere of radius 2.78 it is 2.220 electrons. In other words, more than half of the four electrons in the L shell of the atom are found outside this sphere.

These are the nonoverlapping spheres concerned in our discussion of Chapter 10. It is clear that that discussion is entirely unsuited to this problem in the form we have presented. We can see this further from the fact that outside the nonoverlapping spheres we assumed a constant potential of zero. As a matter of fact, from Table 5-4, we see that the actual potential energy of an electron at the radius $x = 2.46$ is -3.013 rydbergs, whereas with a radius 2.78 it is -2.408 rydbergs. Thus even the lowest of the valence molecular orbitals, the $2\sigma_g$ with an energy of about -2 rydbergs, will have a positive kinetic energy at the radius of the sphere of between 0.4 and 1 rydberg. On the other hand, if we discontinuously shifted the potential outside this radius to zero, the molecular orbital would have to shift discontinuously to a negative kinetic energy of about -2 rydbergs at this point.

The reader should realize that the place where the kinetic energy shifts from positive to negative is the point of inflection on the curve of P as a function of r. This is not just the same as the point of inflection in Fig. 11-2, since the latter figure shows the function R, not $P = Rr$. But in any case, it actually comes at a radius considerably greater than the radius of the nonoverlapping sphere. In other words, the actual atomic potential, which must represent a fairly good first approximation to the potential found in the molecule, would be grossly misrepresented if we used the muffin-tin potential of the type discussed in Chapter 10, using the nonoverlapping spheres, and zero potential outside the spheres. It was not, in fact, assumed in the original treatment of Johnson and Smith that this would be done.

Rather, what they assumed was that outside the nonoverlapping spheres the potential would be constant, but not zero. They assumed a constant, negative value of the potential energy. Thus in a treatment of the C_2 problem by the multiple-scattering method with nonoverlapping spheres, computed by Danese and Connolly using the programs of Johnson and Smith, the potential energy of an electron in this outside region came out to be -1.2934 rydbergs. This particular calculation led to eigenvalues for the various orbitals of -20.3444 rydbergs for $1\sigma_g$, -20.3429 rydbergs for $1\sigma_u$, -1.3495 rydbergs for $2\sigma_g$, -0.6892 rydberg for $2\sigma_u$, -0.6478 rydberg for $1\pi_u$, and -0.4495 rydberg for the unoccupied $3\sigma_g$. We thus see that the kinetic energy of the $2\sigma_g$ in the outer region II with this treatment was very slightly negative, $-1.3495 + 1.2934$ rydbergs $= -0.0561$ rydberg, corresponding to a very slowly decreasing tail; whereas the higher occupied orbitals, $2\sigma_u$ and $1\pi_u$, would have a positive kinetic energy in region II, so that there would be no exponentially decreasing factor in their wave function. In this case one would have to use for their wave functions not the functions $k_l^{(1)}$ of Chapter 10 but the spherical bessel and neumann functions, j_l and n_l, described in detail in Note 4 and going sinusoidally at large distances. The model in this form is entirely unable to explain the exponentially decreasing tail of the wave functions.

To overcome this difficulty, Johnson and Smith introduced the idea of the outer sphere, which we have merely mentioned earlier. This was a larger sphere centered on the midpoint between the atoms and tangent to the nonoverlapping atomic spheres. Thus its radius was twice the radius of the atomic spheres. They assumed a spherical potential energy of an electron outside this outer sphere, which they called region III, going to zero at infinity but gradually decreasing as r decreased to the radius of the outer sphere. Within this region III, all occupied orbitals had negative kinetic energy, and all orbitals were made to go to zero at infinity. The potentials in regions II and III were determined in approximate ways as averages of the correct potentials through these regions of space.

These potentials for the C_2 atom, as given in the computation of Danese and Connolly, are shown in Fig. 11-5. The potential energy of the electron in region I, $-V$, is approximately equal to that of Table 5-4, but lies slightly lower at the radius of the atomic sphere, about $2.5\ x$ units. Then the potential

11 QUALITATIVE NATURE OF MOLECULAR ORBITALS

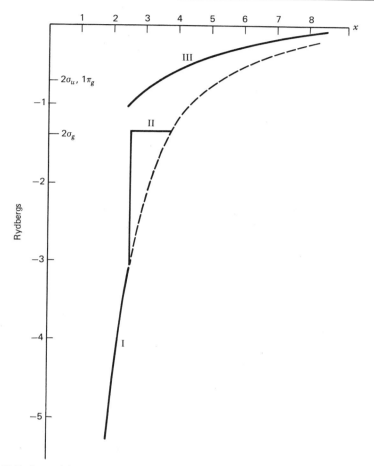

Figure 11-5 Potential energy of an electron, $-V$, as a function of x, distance from nucleus, in C_2 molecule. I, inside nonoverlapping atomic sphere of radius $x = 2.5$. II, constant value in region II. III, value in outer sphere region III as function of distance from midpoint between atoms. Dashed curve, atomic potential joined smoothly to potential in sphere I. One-electron energies of orbitals are indicated.

jumps to the constant value shown in region II. In region III we have a separate curve, not really comparable with the others since the radius x is here measured from the midpoint between the nuclei rather than from either nucleus. The dashed curve is a curve of atomic nature, slightly modified to join smoothly with the potential in region I. We have included it, although it is not concerned in the multiple-scattering calculation, since it must represent quite closely the spherically symmetrical potential for which we are really trying to solve Schrödinger's equation. We have also indicated the one-electron energies of the occupied orbitals.

Quite clearly, we are still doing a good deal of violence to the correct potential by making the muffin-tin approximation and by using the outer sphere, although not nearly as great violence than if the potential were jumping to the value zero for all values of x greater than 2.5. The program of Johnson and Smith for this method regards the outer sphere as another atomic sphere, and it includes the provision for making the wave function and its normal derivative continuous over the surface of the outer sphere. When the output of the program is examined in detail, however, it is found that it does not actually converge well enough to give this continuity on the outer sphere. Thus, if we refer to Fig. 11-3, we can reproduce by the computer output the form of the $2\sigma_g$ and $2\sigma_u$ functions along the internuclear axis. We use the solution for region I from $x = 0$ to 2.5, the nucleus, and again out to $x = 5$, the point of tangency between the region I and III. We find, however, that the computed curve in the outer sphere region III does not join at all smoothly at $x = 5$. Danese and Connolly, who have examined this computer output in more detail, find other areas where there is quite poor fitting. Aside from these rather minor difficulties, the output resembles Fig. 11-3 quite closely, showing that the method is capable of giving fairly reasonable results for the wave functions. In other words, the coefficients C_{lm} of Equation 9-1 are given reasonable values by the computer program.

What Danese and Connolly have done is to start with these computer results and compute from them the charge density and potential throughout space. They have then substituted these quantities into the expression of Equation 2-3 for the total energy and have computed this total energy by laborious numerical means. The net result came out in tolerably good agreement with the Hartree–Fock value and gave an energy for the molecule sufficiently lower than the energy of two isolated atoms to give a failrly adequate account of the atomic binding in the molecule. This leads one to hope that the molecular orbitals computed by this method are fairly good, and that if the calculations could only be made more simply and accurately, they might give a good approach to finding the energy of a molecule in comparison with the atoms from which it was formed.

There are two major drawbacks to this type of calculation. First is a practical one. The geometry of the outer sphere is very awkward to handle, since it disturbs the spherical symmetry one otherwise has around an atom.

11 QUALITATIVE NATURE OF MOLECULAR ORBITALS

This would be overcome if one somehow could eliminate the outer sphere and use entirely a spherically symmetrical potential inside an atomic cell, which in the case of C_2 would be the half space on one side of the plane which is half way between the atoms. Somehow one must take care of the tails of the orbitals and the fact that they go to zero at infinite distance, without the inconvenient and inaccurate approximation of the outer sphere. Second, and even more important, is the inaccuracy introduced by the poor potential used in regions II and III in Fig. 11-5. The dashed curve of that figure without doubt represents the correct potential much better than the curves II and III. In other words, we are led to look for some approach much more closely resembling the cellular method, with a spherical potential.

Johnson and his collaborators have advocated a method of overlapping spheres to get some of these advantages. They have proposed continuing the potential, following something like the dashed curve of Fig. 11-5, out to a radius considerably greater than the radius of the nonoverlapping sphere. For instance, they have suggested a sphere of about twice the volume of the nonoverlapping sphere, hence with a radius $2^{1/3}$ as great, or about 1.26 times as great, or a radius of about $x = 3.2$. This would greatly increase the volume over which the spherical potential was being used and decrease the volume of region II. Using this assumption, they have been able to solve for molecules of very considerable complexity and to get charge densities that seem very reasonable. However, these methods have not removed the computational awkwardness of the procedure, and no reliable calculations of total energy have been made by the method, although approximations have been carried through.

As an illustration of what can be done, we show in Fig. 11-6 the molecule TCNQ, tetracyanoquinodimethane, $C_{12}N_4H_4$, which has been the subject of a great deal of work carried out by Herman and colleagues. It is a planar molecule, which makes it convenient for pedagogic purposes, since we can draw the atoms in a plane. We have shown each atom placed inside an overlapping sphere of approximately the dimensions used by Herman, Johnson, et al. Between any two overlapping atoms we have drawn a plane surface dividing them perpendicular to the plane of the paper, indicating that the spherical potential met in one atom is to be used only on one side of the dividing plane. By extending these planes, we have set up cells, one to an atom. Some of these cells, in particular those of the N atoms, extend out to infinity in the plane of the paper. Others, such as the C atoms, have cells that go to infinity only in a direction at right angles to the plane of the paper.

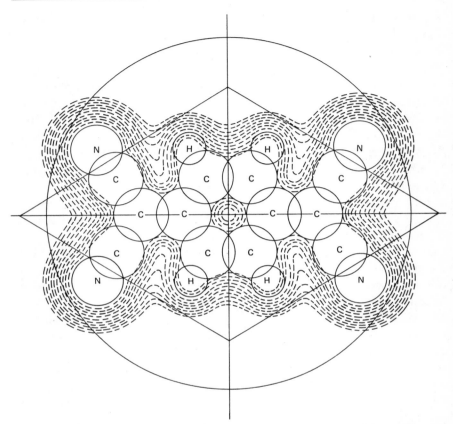

Figure 11-6 TCNQ molecule, showing overlapping spheres and cells. From Johnson, Herman, and Kjaellander, in *Electronic Structure of Polymers and Molecular Crystals Applications of the SCF–Xα Scattered Wave Method to Molecular Crystals and Polymers*, Eds. Andre, Ladik, and Delhalle, Plenum Press, New York, 1975.

Herman and his colleagues have assumed a spherical potential within each spherical cell, a constant potential in the region II outside the cells, and an outer sphere region III outside the large sphere indicated in the figure.

Contours of the density of valence electrons outside the cells are indicated in Fig. 11-6, computed from the multiple-scattering method. They give a remarkably good idea of the nature of the charge density in the region outside the atoms; of course, the calculations also give the density inside the atoms, but it is so high that it is impossible to show it in a diagram like that of the figure. It is the ease of making such calculations with the computer programs developed by Johnson, Herman, and others, combined with the success of Danese and Connolly in computing the total energy of the C_2 molecule, that have given one hope that the general approach has possibilities of development which have not yet been explored. We now go on in the next chapter to possible ways of combining the main features of the multiple-scattering method with a cellular approach to achieve these ends.

12 The cellular method-solution of Schrödinger's equation

From the discussion of the preceding chapter, we can see more clearly the problem we should like to solve: the determination of a solution of Schrödinger's equation in a potential that is spherically symmetrical inside each of the cells of a complicated molecule such as is shown in Fig. 11-6, the potential in the cell having the qualitative behavior of that shown by the dashed curve in Fig. 11-5. The solution should be continuous with continuous derivatives through all space, should be regular at each nucleus, and should be regular at infinity.

We realize, of course, that a solution of a single l value can be regular both at the origin and at infinity only if it is an eigenfunction of the spherical potential. Hence for any other energy parameter than an eigenvalue of the spherical problem, the function of a given l regular at the origin and that of the same l regular at infinity will not be the same. We illustrate this fact in Fig. 12-1. Here we have plotted the P_l function, for the value $l = 1$, for a spherical problem whose potential is that of the dashed curve of Fig. 11-5; the energy parameter is that appropriate to the $2\sigma_u$ molecular orbital. We show the two functions, one regular at the origin, the other regular at infinity. It is obvious that they are different and that at a point where they have the same ordinates, so that the functions join continuously, they have different slopes, so that the first derivative is discontinuous.

We must conclude, then, that if we build up a solution of Schrödinger's equation for the cellular problem out of functions like those of Fig. 12-1,

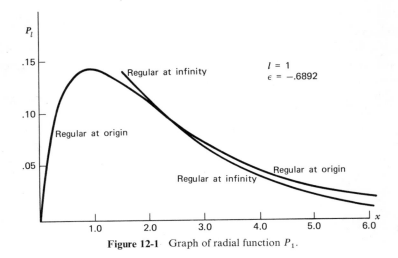

Figure 12-1 Graph of radial function P_1.

multiplied by suitable spherical harmonics of angle, we must be using a sum over functions of different l values. Each of these functions regular at the origin, as we see from Fig. 12-1, will go infinite at infinite r. Hence we must be making up a linear combination of functions of all l values, each going infinite at infinite r, and yet the combination must vanish at infinity. This is an unfamiliar and at first sight an impractical sort of boundary condition. It is, in fact, not difficult to carry out. We can familiarize the reader with the procedure by thinking of a simpler problem, which we have so far avoided: the method of computing the function for a given l value and energy parameter, which is regular at infinity. How, in other words, did we calculate the curves regular at infinity in Fig. 12-1?

This can be done in practice by expanding the solutions of Schrödinger's equation for a given l value in terms of the C and S functions which we have introduced in Chapter 4 in connection with the power series expansion of the functions P_l in powers of $x - x_0$. In Fig. 12-2 we show functions $C(x - x_0)$ and $S(x - x_0)$ for the case $l = 0$, in a potential and with an energy not identical with those used in Fig. 12-1, but quite similar. These functions have been obtained in the following way. First, we carry out a power series expansion in powers of $x - x_0$, following the procedures of Chapter 4. We carry these series out as far as they are adequately convergent. Beyond that point we integrate both inward and outward by Noumerov integration. In this way

12 THE CELLULAR METHOD-SOLUTION OF SCHRÖDINGER'S EQUATION

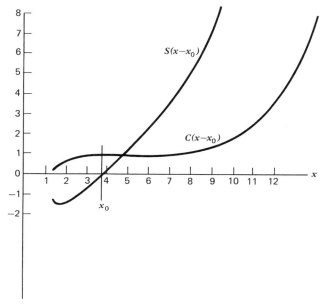

Figure 12-2 Curves $C(x - x_0)$ and $S(x - x_0)$ for carbon atom, $\alpha = 0.77$, $l = 0$, for energy parameter ε of -0.691745 rydberg. The eigenvalue of the $2s$ orbital with this potential is $\varepsilon_{2s} = -0.99831$ rydberg. The value of x_0 is 3.74.

we can extend the functions to as large and small values of x as we please. The results of such a calculation are the curves of Fig. 12-2.

Before we discuss the implications of these curves, there are two matters of technique connected with them. In the first place, Noumerov integration can be carried out both with increasing and decreasing x, but it is more convenient with increasing x. The reason is that, with the Herman–Skillman mesh, points are more closely spaced at small x. If we integrate inward, it is not convenient to decrease the spacing at small x, since we do not have calculations of intermediate points with which to start the integration with closer mesh. Consequently, the integration is only convenient if we use the series expansion to get two points whose spacing is small enough to correspond to the smallest mesh we wish to use, and integrate inward using that mesh. As a practical matter, the calculation for Fig. 12-2 was made $x_0 = 3.74$, and the inward integration was carried out only over the block of points from 1.34 to 3.10, using the interval 0.16.

A second practical point is connected with the outward Noumerov integration. For the cases considered, it is found that the Noumerov method is not sufficiently accurate to use for intervals greater than 0.64. Consequently this interval was used over the range of x out to 19.10, which was as far as the calculation was carried, rather than increasing the intervals to 1.28, as in the Herman–Skillman mesh.

Let us now consider the significance of the curves of Fig. 12-2. An arbitrary linear combination of these two functions will furnish a general solution of Schrödinger's equation for the assumed l and energy parameter. Hence one linear combination $c_1 C + s_1 S$ will give the solution regular at the origin, and another linear combination $c_2 C + s_2 S$ will give the solution regular at infinity. The simplest method of determining c_1 and s_1 is to make the function $c_1 C + s_1 S$ agree with the values of the function regular at the origin at two values of x. We have seen in Chapters 3 and 4 how to get the function regular at the origin by expanding in power series, and then continuing outward by Noumerov integration.

To get the values of c_2 and s_2, we note that both functions C and S can be written as linear combinations of the function regular at the origin and that regular at infinity. The function regular at the origin will be increasing rapidly with x, whereas that regular at infinity will be decreasing. Thus if x is large enough, C and S will each consist almost entirely of the function regular at the origin, so that their ratio as a function of x will approach a constant. We carry our outward Noumerov integration until this ratio has, in fact, become constant to any desired accuracy. Then we can set up a linear combination of the two functions in which the rapidly increasing function is exactly cancelled out, leaving only the function regular at infinity. This procedure in practice is quite simple to use.

We note from Equation 3-6 that at sufficiently large values of r the potential V, as given in Table 5-4, and the term $-l(l + 1)/r^2$ will vanish, leaving only the differential equation $d^2 P_l/dr^2 = -\varepsilon_i P_l$, with solutions

$$P_l = a \exp(\sqrt{-\varepsilon_i} r) + b \exp(-\sqrt{-\varepsilon_i} r) \qquad (12\text{-}1)$$

At sufficiently large r the decreasing exponential has effectively vanished, leaving the function P_l as an increasing exponential. One has to go to a considerably greater value of r for this approximation to be reasonable than for the functions C and S to be proportional to each other. The reason is that the two functions C and S for a given l have the same value of the quantity $-l(l + 1)/r^2$ in Equation 3-6. This term decreases much more slowly than the

12 THE CELLULAR METHOD-SOLUTION OF SCHRÖDINGER'S EQUATION

potential V with increasing r, so that the procedure described in the preceding paragraph need only be carried out to values of r for which V is negligible. On the other hand, for Equation 12-1 to be applicable we must go to the larger r's for which $-l(l+1)/r^2$ is negligible as well. When we do this, however, all P_l's have the same exponential factor, which as we see shortly, make it practical to satisfy the boundary condition that the function goes to zero at infinite r.

We now are sufficiently familiar with the method we shall be using that we can describe it in detail. We do it in terms of the homonuclear diatomic molecule C_2, as before. We express the wave function within one of the cells, which in this case includes the half-space closer to one of the nuclei than to the other, in the form given in Equation 9-1, namely,

$$u(\varepsilon,r,\theta,\phi) = \sum(lm) C_{lm} R_{lm}(\varepsilon,r) Y_{lm}(\theta,\phi) \qquad (12\text{-}2)$$

Here the functions R_{lm} will be chosen to be those regular at the nucleus. We can get them either by integrating all the way out to large r by Noumerov integration, as we did in Table 3-1, or by using series in $x - x_0$ and integrating outward from there, as we have just been describing; both will lead to the same result. Then we demand that the C_{lm}'s should be so chosen that boundary conditions will be satisfied around the boundary of the cell, for certain definite angular directions.

Certain coefficients will automatically be zero, on account of symmetry. The diatomic molecule has rotational symmetry around the line joining the nuclei. Hence the component of angular momentum around the internuclear axis is quantized, and each solution will have a fixed value for the quantum number m. The states with $m = 0$ are σ states, those with $m = \pm 1$ are π states, and the dependence of all functions on the angle ϕ is given by the exponential $\exp(im\phi)$. For the homonuclear molecule C_2, with its reflection symmetry in the plane midway between the nuclei, we can express the boundary conditions across this plane by the statement that a σ_g or π_u state should have zero normal derivative at all points of the plane, whereas a σ_u should have zero value of the function on the plane.

With this particular set of conditions, the dependence of the function on ϕ is determined by m. For a fixed m, let us then choose our angular directions by giving specific values for $\cos\theta$. For example, to be definite, we can choose values of $\cos\theta$ equal to 1 (the direction toward the other nucleus), 0.8, 0.6, 0.4, 0.2, 0, -0.2, -0.4, -0.6, -0.8, -1 (the direction away from the other nucleus). Radii out from the nucleus in the directions from $\cos\theta = -1$ to 0 will go to infinity without meeting another cell. Those with positive $\cos\theta$ will intersect the bounding plane between the atoms.

Along those directions that go to infinity we demand that the wave function go to zero at infinity, as we have been describing in preceding paragraphs. Since all R_{lm}'s approach the same exponential, we can divide out this exponential and get linear equations for the C_{lm}'s in which the radial dependence drops out for large r. Along those directions that intersect the bounding plane, the wave function must go to zero for a σ_u state, or its normal derivative must go to zero for a σ_g or π_u state. Either condition can be written down explicitly in terms of Equation 12-2. In Note 5 we take up the method of taking the normal derivative of the function u, which demands working out the derivative of Y_{lm} with respect to θ. If we use a power series for P_{lm} or R_{lm} in powers of $x - x_0$, we can conveniently compute these functions and their derivatives on the bounding plane.

Since there are 11 directions along which we shall determine boundary conditions, this will lead to 11 simultaneous linear homogeneous equations for the coefficients C_{lm} corresponding to the chosen m value. To satisfy these simultaneous equations, we choose functions with $l = 0, 1, \ldots 10$ for a σ state. The 11 simultaneous homogeneous linear equations for 11 coefficients will have nonvanishing solutions only if the determinant of coefficients is zero. This can be secured by computing the determinant as a function of the energy parameter ε which enters into $R_{lm}(\varepsilon,r)$, and will then lead to 11 eigenvalues, of which the lower ones should represent the occupied σ orbitals. The simultaneous equations will determine the ratios of coefficients, and the remaining degree of freedom is used to normalize the eigenfunction over the whole space consisting of two atoms. Similar assumptions can be made for the π orbitals, where we remember that for $m = 1$, we can have only Y_{lm}'s for $l = 1, 2, \ldots$, and where the wave function is automatically zero for $\cos\theta = \pm 1$.

The resulting series of Equation 12-2 will exactly satisfy boundary conditions of continuity of function and derivative along the 11 directions we have specified. We can well ask, what can we expect for intermediate directions? The answer is that our experience with the success of tenth-order polynomials, as described in Chapter 4, would lead us to expect a very good description of the actual function. The spherical harmonics for l up to 10, with $m = 0$, can be expressed as polynomials, of order up to 10, in $\cos\theta$. Consequently, our expansion of Equation 12-2 is equivalent to a tenth-order polynomial expansion in $\cos\theta$. Tests which have been made of this expansion justify us in this optimism. Of course, there is no reason to expect the expansion to represent any approximation to the actual wave function, outside the cell containing the nucleus around which the expansion is made. In the case of the C_2 molecule, the solution inside this cell will join smoothly onto the identical expansion in the other cell, using the same function for σ_g and π_u, the negative of it for σ_u.

If we compare with our qualitative description of Chapter 11, in terms of the LCAO method, we may expect very similar results to what we have shown in Fig. 11-3. We are not combining functions that go to zero at infinity, as in Fig. 11-2, but rather functions like those of Fig. 12-1, behaving like those of Fig. 11-2 at small r but going infinite at large r. However, the combination of the terms of Equation 12-2 is just such as to make the combined function go to zero at infinity for each of the chosen directions. And the tail of the wave function arising from one nucleus, in the cell containing the other nucleus, is made up out of the terms of higher l value in Equation 12-2, which taken together are able to simulate its form.

Of course, the C_2 molecule is a very simple case. If we had not had a homonuclear molecule—for instance, if we had been discussing CO—we should have had to set up separate expansions in the two cells, with different potentials, one for C, one for O. The boundary conditions over the bounding plane between the two atoms would have to be set up to match the function and its normal derivative on the two sides of the plane. We would have to have twice as many undetermined coefficients, and twice as big a secular equation. Problems of just this kind are met in the multiple-scattering method, and experience there can well be a guide to what to expect with the present method. In particular, it is found that a secular equation considerably smaller than we are contemplating is enough to give quite a good value for the eigenvalues. Then one can go on to solve the whole set of equations to get the coefficients required for a good expansion of the wave function. This sort of thing is, in fact, done in setting up such density contours as are shown in Fig. 11-6. There is no reason to believe that the present method will lead to much more elaborate calculations than are used there.

Once we have solved Schrödinger's equation for a molecular potential, we must realize that we have only made the first step toward self-consistency. And our spherically symmetric potential is not general enough to describe the actual potential we find inside an atomic cell in the molecule. But at least, as Danese and Connolly have shown, we can go a long way on the basis of such solutions for the molecular orbitals as we are describing in this chapter. We shall now go on to consider the further steps toward self-consistency.

13 Self-consistent field for molecules

Once we have determined unnormalized molecular orbitals, as described in the preceding chapter, we must carry out the same sort of steps to find a self-consistent field which we have described for atoms in Chapter 5. In this process we need to carry out volume integrations over the volume of the cell, the integrand being in the form of a sum of functions, each consisting of a function of r multiplied by a function of angles. We first consider how to carry out these integrations. We take up explicitly only the case such as we meet in C_2, where the function is independent of the angle ϕ. Our general approach is to carry out the integration over θ analytically for each value of r and then to integrate numerically over r, using Simpson's rule with the Herman–Skillman mesh, as in Chapter 5.

Let r_0 be the distance from the nucleus to the plane bounding the cell, which in the case of C_2 is the midplane between atoms. We use spherical coordinates with the z axis pointing from the nucleus in the cell toward the other nucleus. We use $w = \cos\theta$ as a variable, rather than θ itself. Then for a given value of r greater than r_0, the range of w lying inside the cell is from $w = -1$ to r_0/r. Inside the nonoverlapping atomic sphere, for r less than r_0, the range of w is -1 to 1. We express the angular dependence of the function to be integrated as a tenth-degree polynomial in w. We do this by fitting the function exactly at 11 equally spaced points along the w axis, using the method of Note 2. Outside the atomic sphere, these points should all be inside the range from $w = -1$ to r_0/r, where our solution for the molecular orbital is valid. Ordinarily they would be chosen to be the two end points -1 and r_0/r, with equally spaced intermediate points. Inside the atomic sphere they would be the points from 1 to -1, which we have already used in the preceding chapter to define the directions on which we would satisfy boundary conditions from one cell to the other, or at infinity. The integration of the polynomial is trivial.

We must be able, then, to compute the numerical values of the function to be integrated at 11 points representing different values of w, for each r used in the Herman–Skillman mesh. This is no major computational task. In Note 6 we give formulas for the Y_{lm}'s for l up to 10, in terms of polynomials in w, and also the inverse formulas for w^n in terms of the spherical harmonics. This would allow us to compute the wave functions of Equation 12-2 at the various

13 SELF-CONSISTENT FIELD FOR MOLECULES

points where we need their values. When we square the wave functions to get the charge density, we do not try to square analytically the series of Equation 12-2, but rather we simply square each of the numerical entries involved in the table representing the function.

First, for normalization, we merely use the squares of the wave functions as computed from Equation 12-2. We can equally well multiply two wave functions together, to check the orthogonality of the wave functions. Since the wave functions are solutions of a Schrödinger equation, with the spherically symmetrical potential inside each cell, our general proof of orthogonality of solutions of Schrödinger's equations is valid, and any departures from orthogonality would be only an indication of the degree of accuracy of the calculations. Of course, symmetry would automatically bring about orthogonality except between two orbitals of the same symmetry, namely, between the $1\sigma_g$ and $2\sigma_g$ or between the $1\sigma_u$ and the $2\sigma_u$. These orthogonalities, which have given so much trouble with the LCAO method, will automatically be brought about.

After the orbitals are normalized, we must find the potential V, as in Chapters 5 and 6. First we consider the electrostatic potential arising from the charge density. The nuclear potential $\sum(a) 2Z_a/r_{1a}$ is obvious. For the potential arising from all electronic charge, $V_e(1)$, we use a generalization of Equation 5-5, holding for nonspherical charge distributions. We assume that the total charge distribution is expressed as a series

$$\rho = \sum(lm)\rho_{lm}(r) Y_{lm}(\theta,\phi) \tag{13-1}$$

Then this generalization is

$$V_e(1) = \sum(lm) Y_{lm}(\theta,\phi) \frac{4\pi}{2l+1} \left[\frac{2}{r_1^{l+1}} \int_0^{r_1} r_2^{l+2} \rho_{lm}(r_2) \, dr_2 \right.$$
$$\left. + r_1^l \int_{r_1}^{\infty} 2r_2^{-l+1} \rho_{lm}(r_2) \, dr_2 \right] \tag{13-2}$$

The quantities ρ_{lm} are determined by

$$\rho_{lm}(r) = \int \rho(r,\theta,\phi) Y_{lm}^*(\theta,\phi) \, d\Omega \tag{13-3}$$

where the integration is over all angles. When we have a charge density independent of angles, all ρ_{lm}'s are zero except that for $l = m = 0$. The quantity Y_{00} equals $(4\pi)^{-1/2}$, as we point out in Note 4. The integral of unity over all solid angles is 4π. Consequently, we have $\rho_{00}(r) = (4\pi)^{1/2}\rho(r)$, and $Y_{00}\rho_{00}(r)$, which appears in Equation 13-2, equals $\rho(r)$. Consequently for this case Equation 13-2 reduces to Equation 5-5.

We divide the problem of finding the quantity V_e into that of finding the contributions of each separate atomic cell, and then we add these separate potentials to get the total. For the origin of the expansion of the potential given in Equation 13-2, we naturally use the nucleus of the atom in the cell. Then Equation 13-2 shows the potential as arising from two types of terms. The first term, in the bracket in Equation 13-2, is in $r_1^{-(l+1)}$, infinite at the origin but regular at infinity. When we take account of the integral involved in this term, however, this makes the complete term finite everywhere. It is the potential of a multipole located at the origin. For $l = 0$, the integral is $\int_0^{r_1} 4\pi r_2^2 \rho(r_2)\,dr_2$, and it is simply the total charge inside a sphere of radius r_1, as we pointed out in the discussion of Equation 5-5. The potential at a point distant r_1 from the origin is the potential of this charge, as if it were all concentrated at the origin. The second term in this case is the outer shielding term, arising from all charge in the cell located at values of r outside r_1. If we go to a value of r_1 outside all points of the cell, we have only the $1/r_1$ term left, so that the potential of the charge in the cell is just what would arise from the whole electronic charge in the cell, concentrated at the origin. If the atom in the cell is electrically neutral, this will be exactly balanced by the potential arising from the nucleus.

The next term in Equation 13-2, for $l = 1$, is at large distances the potential of a dipole arising from the total dipole moment of the electronic charge in the cell. For $l = 2$ we have the quadrupole moment and so on. Hence the potential of the whole charge in the cell, at a distant point, will be the sum of potentials of the total charge and of all the various multipoles located at the origin of the atom. We must remember, however, that if we are at distances r_1 small enough so that appreciable charge in the cell is located at larger distance than this, there will be modifications in this potential arising from outer shielding.

We wish to use these expressions for potential to compute the actual potential inside a particular cell, say, that around atom a, arising not only from the charge in cell a but also from that in all other cells. We can get a fairly good approximation to this potential by simply using the long-distance expansion in terms of multipoles, which we have described in the preceding paragraph. For more accurate work, we should have to use the whole of expression 13-2, to get the correct value of the potential exerted by a nearby

13 SELF-CONSISTENT FIELD FOR MOLECULES

cell b on a given cell a. In any case, however, Equation 13-2 involves only integrals over r_2, over a cell, of simple functions of r_2, a type of integral we have already been calculating in Chapters 5 and 6. Thus the calculation of the total potential arising from electrostatics $V_N + V_e$ inside a given cell can be rigorously obtained without difficulty.

It is to be noted that the contributions of the higher multipoles or higher l values from distant cells will fall off rapidly on account of the term $r_1^{-(l+1)}$, so that the series in l will be rapidly convergent. The term arising from the sum of the potentials of the total charge in each cell, the term for $l = 0$, is the familiar Madelung potential, which would be found by familiar methods. The term for $l = 1$ is the potential arising from dipoles distributed throughout the molecule, familiar from dielectric constant calculations. Both of these terms, for $l = 0$ and 1, depend on the details of the shape and size of the molecule, since they have large contributions from the surface behavior of the system. The terms of higher l value converge rapidly enough as we go to more and more distant atoms, so that they are largely independent of the shape and size of the molecule.

In addition to these purely electrostatic effects, we have the term in $\rho^{-1/3}$, the $V_{X\alpha}$ term. This is merely a local term, not involving an integration as the coulomb potential does. Since we have already assumed that the charge density is available at each value of r and w involved in our mesh of points, we simply take the $1/3$ power of each of these values to get the local exchange potential. We can then use Equation 13-3 to get the coefficients of the expansion of the exchange potential in the form of Equation 13-1, in terms of spherical harmonics of angles and radial functions.

We are thus able to set up an expansion of the quantity $-(\varepsilon + V)$ of Equations 6-1 and 6-2, in the form

$$-(\varepsilon + V) = \sum (lm) V_{lm}(r) Y_{lm}(\theta,\phi) \qquad (13\text{-}4)$$

We should properly be using this potential as the starting point of a next solution of Schrödinger's equation, in an iterative procedure to get an exact solution of the self-consistent field for the molecule. However, we have discussed Schrödinger's equation only for a spherical potential. Hence we use only the term for $l = m = 0$ in Equation 13-4 as the spherical potential for the approximation we have so far considered, in which the wave function is a solution of a spherical potential problem within each cell. It is now a straightforward procedure to carry through this sort of self-consistency, which is all that has been contemplated in calculations made up to date. In other words, we demand that the solutions of Equation 9-1 for the spherical potential problem should lead, through the procedures we have been describing, to a potential like Equation 13-4, whose term for $l = 0$ is identical with the originally assumed potential.

We are able, however, to go one step further in the determination of the total energy, as Danese and Connolly have shown. Our charge density of Equation 13-3, and our potential of Equation 13-4, have angular dependences that represent quite a good approximation to the expected exact values. We can then use these expressions in computing the type of integrals over the cell we have described for the isolated atom in Table 6-1 and in Chapter 6 in general. Each of these integrals involves an integration over the cell of the type we have been discussing in this chapter. It is not essentially more difficult to find any one of these integrals, once we have determined the quantities ρ, V_N, V_e, and V_X, than it is to compute one of the normalization integrals. Consequently, we have reduced the calculation of Danese and Connolly, which was carried through by very elaborate numerical integrations, to a set of one-dimensional numerical integrations which can conveniently be carried out by Simpson's rule.

It is probable that this degree of sophistication is as far as one will need to carry the procedure for the present. However, as we show in the next chapter, one can at least formulate the further step of solving Schrödinger's equation, not merely in the spherically symmetrical field, but in the nonspherical potential of Equation 13-4. This, in principle, would allow us to solve the self-consistent field method exactly for the $X\alpha$ method, although in practice it seems likely that it would be so time-consuming as to be impractical.

14 The nonspherical potential

We now consider the general case of the solution of Schrödinger's equation in the nonspherical potential given in Equation 13-4. As with the spherical potential, the general method we use is a power series expansion around the

14 THE NONSPHERICAL POTENTIAL

nucleus, followed by a Noumerov integration beyond the small values of r for which the power series converges. We could also develop power series around other radii r_0, although we do not go into the details of this procedure. We first consider the Noumerov integration.

The characteristic of this problem is that for any exact solution we must use a sum of terms with different l and m values, instead of limiting ourselves to a single l value as with the spherical potential. In other words, something like a hybridization is automatically produced by the nonspherical potential. This is familiar from the theory of the dielectric constant of an isolated atom. In this case the nonspherical potential reduces to a term proportional to the distance along the axis in the direction of the field, or proportional to $r \cos \theta$, a term found for $l = 1$. It is well known that the effect of this perturbation is to mix a little of a p orbital, $l = 1$, in with an s orbital, $l = 0$, to produce the polarized orbital, with a dipole moment proportional to the external field. More generally, a potential like that of Equation 13-4 will produce a mixture of orbitals of all l values.

Let us then start by assuming that the solution for the function $P = ru$ is given by a sum,

$$P = \sum (lm) p_{lm}(r) Y_{lm}(\theta, \phi) \qquad (14\text{-}1)$$

where the p_{lm}'s are functions of r. We wish to solve Equation 3-1, $\nabla^2 u = -(\varepsilon + V)u_i$, where $u = P/r$ and where the potential is given by Equation 13-4, $-(\varepsilon + V) = \sum (lm) v_{lm}(r) Y_{lm}(\theta, \phi)$. We use the methods of Equations 3-2 to 3-6, and show that

$$\sum (lm) \left[\frac{d^2 p_{lm}}{dr^2} - \frac{l(l+1)}{r^2} p_{lm} \right] Y_{lm} = \sum (l'm'l''m'') v_{l'm'} p_{l''m''} Y_{l'm'} Y_{l''m''} \qquad (14\text{-}2)$$

We multiply this equation by a particular Y_{lm}^*, integrate over all solid angles, and take advantage of the fact that the integral of any Y_{lm}^* times another of the Y_{lm}'s over all solid angles is unity if the l's and m's are the same, zero otherwise. Thus we find

$$\frac{d^2 p_{lm}}{dr^2} - \frac{l(l+1)}{r^2} p_{lm} = \sum (l'm'l''m'') v_{l'm'} p_{l''m''} \int Y_{lm}^* Y_{l'm'} Y_{l''m''} d\Omega \qquad (14\text{-}3)$$

The term for $l' = m' = 0$, for the spherical potential, can be handled separately. Here we have the integral of the product of only two spherical harmonics, since $Y_{00} = (4\pi)^{-1/2}$. Thus we use the properties of these integrals which have just been stated, and we see that the resulting term on the right side is $-\varepsilon$ minus the spherical potential. Thus Equation 14-3 is reduced to

$$\frac{d^2 p_{lm}}{dr^2} - \frac{l(l+1)}{r^2} p_{lm} = -(\varepsilon + \text{spherical potential})p_{lm}$$

$$+ \sum{}'(l'm'l''m'')v_{l'm'}p_{l''m''} \int Y_{lm}^* Y_{l'm'} Y_{l''m''} \, d\Omega \quad (14\text{-}4)$$

where $\sum'$ indicates that the term v_{00} is omitted. It is only the last double sum that contains the nonspherical potential. The remaining terms are identical with those of Equations 3-6 and 3-7.

The integral of the product of three spherical harmonics over all solid angles is one we have met before, in Equation 10-19. If we replace l'' by L, and m'' by $m - m'$ (the only case in which the integral is nonvanishing), the integral over solid angles reduces to

$$\int Y_{L,m-m'} Y_{lm}^* Y_{l'm'} \, d\Omega = I_L(lm;l'm') = \sqrt{\frac{2L+1}{4\pi}} \, c^L(lm;l'm') \quad (14\text{-}5)$$

Here the notation $I_L(lm;l'm')$ was used in Equation 10-19. The quantity $c^L(lm;l'm')$ occurs in the theory of atomic multiplets, and the relation between this quantity and the integral of the product of three spherical harmonics is easily proved, as shown in Note 6. Tables of these $c^L(lm;l'm')$ coefficients are given in books on multiplets. Computer programs, described in Note 6, exist for computing the $I_L(lm;l'm')$ coefficients for any values of the indices. The vector-coupling or triangle rule shows that the coefficients are different from zero only for $L = l + l', l + l' - 2, l + l' - 4, \ldots |l - l'|$.

When we make the change in notation given in Equation 14-5, Equation 14-4 is transformed into

$$\frac{d^2 p_{lm}}{dr^2} = -\left[\varepsilon + \text{spherical potential} - \frac{l(l+1)}{r^2}\right] p_{lm} \quad (14\text{-}6)$$

$$+ \sum{}'(Ll'm')I_L(lm;l'm')v_{l'm'}p_{L,m-m'} \quad (14\text{-}6)$$

where as before $\sum'$ indicates that the term v_{00} is omitted. These differential equations of Equation 14-6 are generally called coupled differential equations, since each one couples all the radial functions p_{lm}. It is simple to set up methods of solving them, but it is generally considered that the labor, and particularly the problem of storage of information involved in them, is prohibitively complicated. However, we may as well write down the

14 THE NONSPHERICAL POTENTIAL

procedure. To make the notation more compact, let us rewrite Equation 14-6 in the form

$$\frac{d^2 p_{lm}}{dr^2} = g p_{lm} + G_{lm} \tag{14-7}$$

First we consider the Noumerov integration. We introduce a quantity y_n related to p_{lmn} (where in this paragraph n refers to the entry in a table of p_{lm}) by Equation 3-11. That is, we write

$$y_n = (p_{lm})_n (1 - g_n h^2/12) - (h^2/12)(G_{lm})_n \tag{14-8}$$

We substitute this into Equation 3-13, and we have

$$y_{n+1} = \left(2 + \frac{g_n h^2}{1 - g_n h^2/12}\right) y_n - y_{n-1} + \frac{h^2 (G_{lm})_n}{1 - g_n h^2/12} \tag{14-9}$$

This equation is as convenient as the ordinary Noumerov equation of Equation 3-15 to use, aside from the labor of computing the terms G_{lm}.

We must make as many simultaneous Noumerov integrations as we retain terms in Equation 13-4. For a complete solution, as we have pointed out, this is almost prohibitively complicated. However, in actual cases, the number of terms required may be greatly reduced. We have noticed that the term in $\cos \theta$, corresponding to $l' = 1, m' = 0$, gives the effect of a constant external field. In the case of C_2, this term is probably all that would be required to give a fairly good idea of the effect of a nonspherical potential on the molecular orbitals. In this case, on account of the vector coupling rule, we should be able to get a reasonable first approximation by mixing only a p function with our s function, or only an s and d with a p function. In other words, we could arbitrarily exclude all contributions aside from these and probably get usable results. This would only multiply the labor by a factor of two or three.

Next we consider the power series expansion of the problem in powers of r, needed to start the Noumerov integration from the origin. First, the functions $v_{lm}(r)$ of Equation 13-4, in the expansion of $-(\varepsilon + V)$, must be expanded as power series in r. Thus let

$$v_{lm}(r) = \sum (n) v_{lmn} r^n \tag{14-10}$$

where in this discussion n refers to the power of r. This expansion is discussed in Note 7, where we show that for each l value, and aside from the nuclear term in r^{-1}, the lowest value of n in the expansion is $n = l$. Similarly, we must expand the functions p_{lm} of Equation 14-1 in power series,

$$p_{lm}(r) = \sum(n) p_{lmn} r^n \qquad (14\text{-}11)$$

We now substitute these expressions into Equation 14-6, and follow the same line of argument as in Equations 4-6 to 4-12. We include the nuclear potential $-2Z/r$ as the term for $n = -1$ in Equation 14-10, and we find as the generalized form of Equation 4-10 the following:

$$[n(n-1) - l(l+1)] p_{lmn} = \sum (Ll'm'n') I_L(lm;l'n') v_{l'm'n'} p_{L,m-m',n-n'-2} \qquad (14\text{-}12)$$

In starting the use of Equation 14-12, it is useful to think of the expression

$$P = \sum (lmn) p_{lmn} r^n Y_{lm} \qquad (14\text{-}13)$$

as expanding the coefficient of r^n as a function of angles, $\sum(lm) p_{lmn} Y_{lm}$. We shall have independent solutions of Schrödinger's equation for a given energy ε corresponding to each set of values l, m. In general, with a nonspherical potential, the solutions for the same l value but different m will differ from each other, rather than having identical functions of r as with the spherical potential. For one of these independent solutions, we must start our iteration with the term for which $n = l + 1$, with no terms of smaller n values. In this case the factor $n(n-1) - l(l+1) = 0$, the right-hand side of Equation 14-12 is zero, and the equation is satisfied for arbitrary pl, m, $l + 1$. This is the same starting point which we have for the spherical potential. But when we get to higher n values, we find that eventually terms of different l and m values come into the problem.

As an example of what we find, let us consider the simplest case, in which only the nonspherical terms for $l = 1$, $m = 0$ are different from zero. Let us take an s state, for which the leading term, as considered above, has $l = 0$, $n = 1$. Then we find that the terms for $n = 2$ and 3 are just as for the spherical potential. The effect of the nonspherical potential first comes for $n = 4$ and all higher powers of r. Similarly, the effect of $l = 2$ terms in the potential is first felt in the $n = 5$ terms in the wave function P and so on. This first nonvanishing term p_{104} for $l = 1$, $n = 4$, is given by Equation 14-12 as

$$(4.3 - 1.2) p_{104} = \sum(L) I_L(10;10) v_{101} p_{L01} \qquad (14\text{-}14)$$

The only value of L giving a nonvanishing $I_L(10;10)$ is $L = 0$, for which it is $(4\pi)^{-1/2}$. Thus we have

$$10 p_{104} = v_{101} p_{001} (4\pi)^{-1/2} \qquad (14\text{-}15)$$

14 THE NONSPHERICAL POTENTIAL

Similarly

$$18\, p_{105} = (v_{101}p_{002} + v_{102}p_{001})(4\pi)^{-1/2}$$
$$14\, p_{205} = v_{202}p_{001}(4\pi)^{-1/2} \qquad (14\text{-}16)$$

The relations for the other l values are equally simple.

It is clear from this that the higher-order multipole terms in the potential will only have an effect on higher powers of r and that they will have small effect for small r. Thus if we only wish to use our power series to start the Noumerov integration, as by computing entries for $x = 0.01$ and 0.02, in the Herman–Skillman mesh, we shall not have to compute the power series for very many terms. We can then start the Noumerov integration of Equation 14-9 rather easily, and our problem is simply to compute the quantities G_{lm}, the last terms in Equation 14-6, for as many values of l', m', L, and m as are necessary. As we have mentioned earlier, this is a major computational effort, but it should not be out of the question.

After we have found these functions, solutions of the Schrödinger equation in the nonspherical potential corresponding to a given energy ε and the various values of l and m, we must build up a solution like those of Equations 9-1 and 12-2, arbitrary linear combinations of these independent solutions of the problem in the nonspherical potential, to give us a solution of the molecular orbital. Each of these functions will have contributions from all l and m values, but the linear combination can still be written as a linear combination of functions of r times spherical harmonics of angles. There will be no more arbitrary constants C_{lm} in the final solution than we had in Equation 12-2. We can still apply precisely the same boundary conditions over the surface of the cell as with the spherically symmetrical potential problem of Chapter 12. And when we have finally solved for the molecular orbitals, they will be no more complicated than before, still linear combinations of products of functions of r times spherical harmonics of angle. Thus the problem of calculating energy integrals, as described in Chapter 13, is precisely as in the spherically symmetrical potential case. In other words, we have described a procedure by which it is in principle possible to get a really exact solution of Equation 2-4, the one-electron problem of the $X\alpha$ self-consistent field method, and from that to get the total energy of Equation 2-3.

NOTES

The following notes were not included in Professor Slater's original manuscript. They have been inferred from references in the test.

Note 1: The quantities for atomic orbitals $[P(x)]$ and potentials $[U(x)]$ for the carbon atom may be found in Chapter 6 of Herman and Skillman's *Atomic Structure Calculations*, Prentice Hall, Englewood Cliffs, N.J., 1963.

Note 2: The methods for expansion of functions in tenth-order polynomials can be found in "Power Series Methods for Cellular Calculations on Atoms, Molecules and Solids" by J. C. Slater, in *Quantum Science*, J. L. Calais, O. Goscinski, J. Linderberg, and Y. Öhrn, Eds., Plenum Press, New York, 1976, p. 215.

Note 3: For a discussion of variational methods see Chapter 1 of J. C. Slater, *The Self Consistent Field for Molecules and Solids: Quantum Theory of Molecules and Solids*, Vol. 4, McGraw-Hill, New York, 1974.

Note 4: For expansion of plane waves in spherical bessel and spherical neumann functions, see Chapter 9 of J. C. Slater *The Self Consistent Field for Molecules and Solids: Quantum Theory of Molecules and Solids*, Vol. 2, McGraw-Hill, New York, 1965. Prof. Slater evidently intended to provide an appendix with the expansion in spherical harmonics, but the appendix is not extant.

Note 5: See J. C. Slater and J. W. D. Connolly, *Int. J. Quantum Chem.* **10S**, 141 (1976).

Note 6: The formulas referred to may be easily generated from general formulas for spherical harmonics such as found in J. C. Slater, *Quantum Theory of Matter*, McGraw-Hill, New York, 1968. Apparently, Prof. Slater intended to display these in an appendix.

Note 7: Expansion of a potential in powers of r may be found on p. 220 in the reference given in Note 2.

BIBLIOGRAPHY

The following list of books and papers were not included in Professor Slater's original manuscript. They have been inferred from references to them in the text. For the interested reader, the best introduction to this book can be found in Professor Slater's previous works:

Quantum Theory of Atomic Structure

Volumes 1 and 2, McGraw-Hill, New York, 1960.

Quantum Theory of Molecules and Solids

Volume 1, Electronic Structure of Molecules, McGraw-Hill, New York, 1963.
Volume 2, Symmetry and Energy Bands in Crystals, McGraw-Hill, New York, 1965.
Volume 3, Insulators, Semiconductors and Metals, McGraw-Hill, New York, 1967.
Volume 4, The Self-Consistent Field for Molecules and Solids, McGraw-Hill, New York, 1974.

Quantum Theory of Matter, 2nd ed., McGraw-Hill, New York. 1968.

Solid State and Molecular Theory: A Scientific Biography, Wiley, New York, 1975.

The rest of this list consists of those articles referred to in each chapter of the manuscript:

Chapter 1

D. R. Hartree, *The Calculation of Atomic Structures* (John Wiley, New York 1958).
D. R. Hartree, *Proc. Camb. Phil. Soc.* **24**, 89, 111, 426 (1928).
F. Hund, *Z. Physik* **40**, 742; **42**, 93: **43**, 805 (1927).
R. S. Mulliken, *Phys. Rev.* **32**, 186 (1928).
J. E. Lennard-Jones, *Trans. Faraday Soc.* **25**, 668 (1929).
W. Heisenberg, *Z. Physik* **49**, 619 (1928).
F. Bloch, *Z. Physik* **52**, 555 (1928).
L. Brillouin, *Compt. Rend.* **191**, 198 (1930); *J. Phys. Radium* **1**, 377 (1930).
A. H. Wilson, *Proc. Roy. Soc. (Lond.)* **A133**, 458 (1931); **A134**, 277 (1931); **A138**, 594 (1932)
C. C. J. Roothaan, *Rev. Mod. Phys.* **23**, 69 (1951).
S. F. Boys, *Proc. Roy. Soc. (Lond.)* **A200**, 542 (1950).
E. Wigner and E. Seitz, *Phys. Rev.* **43**, 804 (1934); **46**, 509 (1934).
P. A. M. Dirac, *Proc. Camb. Phil. Soc.* **26**, 376 (1930).
K. H. Johnson, *J. Chem. Phys.* **45**, 3085 (1966).
K. H. Johnson and F. C. Smith, Jr., *Phys. Rev. Lett.* **24**, 139 (1970).

Chapter 2

J. C. Slater, *Phys. Rev.* **81**, 385 (1951).
R. Gaspar, *Acta Phys. Akad. Sci. Hung.* **3**, 263 (1954).
W. Kohn and L. J. Sham, *Phys. Rev.* **140**, A1133 (1965).
K. Schwarz, *Phys. Rev. B* **5**, 2466 (1972).
K. Schwarz and J. W. D. Connolly, *J. Chem. Phys.* **55**, 4710 (1971).

Chapter 3

F. Herman and S. Skillman, *Atomic Structure Calculations*, Prentice-Hall, Englewood Cliffs, N. J. 1963.
R. Latter, *Phys. Rev.* **99**, 510 (1955).

Chapter 6

J. B. Mann, *Los Alamos Sci. Lab. Rept. LA-3690 and LA-3691* (1968).

Chapter 8

C. E. Moore, *Atomic Energy Levels*, N. B. S. Circular #467 (USGPO, Washington, D.C. 1949).

Chapter 9

J. C. Slater, *Phys. Rev.* **35**, 509 (1930); *Rev. Mod. Phys.* **6**, 209 (1934).
F. von der Lage and H. Bethe, *Phys. Rev.* **71**, 612 (1947).
J. C. Slater, *Phys. Rev.* **51**, 846 (1937).
J. Korringa, *Physica* **13**, 392 (1947).
W. Kohn and J. Rostoker, *Phys. Rev.* **94**, 1111 (1954).
B. Segall, *Phys. Rev.* **125**, 109 (1962).
G. Burdick, *Phys. Rev.* **129**, 138 (1963).

Chapter 11

J. B. Danese and J. W. D. Connolly, *J. Chem. Phys.* **61**, 3063 (1974).
B. J. Ransil, *Rev. Mod. Phys.* **31**, 239, 245 (1960).
F. Herman, A. R. Williams and K. H. Johnson, *J. Chem. Phys.* **61**, 3508 (1974).
K. H. Johnson, F. Herman and R. Kjællander, in *Electronic Structure of Polymers and Molecular Crystals*, Eds. J. M. Andre, J. Ladik, and J. Delhalle, Plenum Press, New York, 1975, p. 601.

Index

a, 11
Antisymmetry, 5, 9
APW, 60, 61
Atomic orbitals, 1, 44, 72
Augmented plane wave, 59

Basis sets, 7
Block condition, 57
Bohr, 2
Boys, S. F., 7

C_2, 70, 76, 77, 79, 85, 87, 88, 95
Carbon, 30, 33, 36, 42, 44, 45, 50, 52, 54, 83
    2s orbital, 28
Carbon atom, 35, 38, 48
Cell, 56
Cellular method, 55, 56, 58, 81
Charge, 38
Charge density, 2, 38, 39
Clementi, E., 7
CO, 87
Connolly, J. W. D., 12, 24, 27, 47, 76, 78, 87, 92
Convergence, 27
Correlation effect, 10
Correspondence principle, 53

Density, 9
Determinant, 5
Double-zeta, 12

Empty lattice test, 58
Energy band, 58
Energy band theory, 7
Energy term, 10

Exchange, 6, 7
Exchange potential, 39
Exchange term, 8, 11, 53
Excitation, 49
Excitation energy, 53
Excitation state, 53

Fermi hole, 9, 10
Free-electron gas, 11

Gaussian functions, 7

Hamiltonian, 4
Hartree, D. R., 2
Hartree-Fock equations, 6
Hartree-Fock method, 6
Hartree-Fock total energy, 12, 46
Hartree unit, 2
Herman, 79
Herman-Skillman, 16, 20, 21, 23, 24, 33, 34, 35, 39, 45
Herman-Skillman integration mesh, 20
Herman-Skillman mesh, 16, 29, 37, 88
Homonuclear diatomic energies, 71
Hybridization, 72, 93
Hybrid orbitals, 73
Hyper-Hartree-Fock, 48

Ionization energies, 51, 52
Iteration, 3

Johnson, K. H., 8, 62, 79
Johnson and Smith, 61, 63, 76

Kinetic energy, 44, 45, 46
Kohn-Sham, 11

Koopmans' theorem, 49, 52
Korringa, Kohn and Rostoker (KKR), 60, 61, 68, 69

Lage, von der, and Bethe, 58
Lagrangian interpolation, 34
Latter correction, 16, 41, 45
LCAO, 6, 12, 69, 87

Madelung potential, 91
Mann, J. B., 46, 48
Molecular orbitals, 1, 69
Molecular orbital theory, 70
MS-Xα, 8
Muffin-tin, 78
Muffin-Tin Method, 55
Muffin-tin potential, 59, 76
Multiple-scattering method, 63, 69, 80
Multiple-scattering treatment, (MS-Xα), 60
Multiplets, 54

Non-muffin-tin corrections, effect of, 60
Nonspherical potential, 92, 94, 97
Noumerov expansion, 20
Noumerov integration, 15, 16, 17, 32, 33, 34, 82, 83, 84, 85, 93, 95, 97

One-electron energies, 50, 71
Orbitals, 1
Outer sphere, 76, 78
Overlapping-sphere, 62, 79

Pauli exclusion principle, 5
Plane wave, 8, 59
Polynomial, 23, 29
Potential, 42
Potential energy, 46
Power series, 20, 21, 22, 24, 29, 32

Radial charge density, 38
Ransil, B. J., 72
Rydberg unit, 2
Roothaan, C. C. J., 6

Scattered-wave method, 68, 69
Schrödinger equation, 3, 6, 11, 13, 44
Schwarz, K., 12, 46, 47, 48
Seitz, F., 56
Self-consistent field, 1, 3, 35
Self-consistent field method, 13
Self-consistent potential, 39
Simpson, 36
Simpson's rule, 35, 37, 88
Smith, F. C., Jr., 8
Spherical coordinates, 13
Spherical harmonics, 13
Spin orbital, 5, 6
Spin polarization, 53, 54

Tenth-degree polynomial, 88
Tetracyanoquinodimethane (TCNQ), 79, 80
Thomas Fermi model, 16
Total energy, 5, 11, 43, 44, 46
Transition state, 49, 52, 53

Unrestricted Hartree-Fock, 53
Unrestricted self-consistent field, 53

$V_{X\alpha}$, 11
Variation method, 46
Virial theorem, 12, 47

Wave vector, 8
Wigner, E., 56
Wigner-Seitz cell, 58
Wronskian, 67

Xα, 8

RETURN TO: CHEMISTRY LIBRARY

100 Hildebrand Hall • 510-642-3753

LOAN PERIOD	1	2 *1 Month*	3
4		5	6

ALL BOOKS MAY BE RECALLED AFTER 7 DAYS.

Renewals may be requested by phone ~~or, using GLADIS, type~~ **inv** ~~followed by your patron ID number.~~

DUE AS STAMPED BELOW.

~~MAY 23~~		

FORM NO. DD 10
3M 7-08

UNIVERSITY OF CALIFORNIA, BERKELEY
Berkeley, California 94720–6000